T0335805

This book is based on a series of lectures delivered under the auspices of the University of Siena. The contents reflect and delineate the author's biochemical research. In particular, the book focuses on the special role played by bound carbohydrates in nature. The complexity of these enigmatic compounds is examined through special case studies including glycoproteins and the metabolism of sialic acids. This book will be of interest, not only to those working in biochemistry and molecular biology, but also to those in the pharmaceutical industry. It provides an entertaining and first hand account for researchers and graduate students.

Bound carbohydrates in nature

Lezioni Lincee
Sponsored by *Foundazione IBM Italia*
Editor: Luigi A. Radicati di Brozolo, Scuola Normale Superiore, Pisa

This series of books arises from lectures given under the auspices of
the Accademia Nazionale dei Lincei and is sponsored by
Foundazione IBM Italia.
The lectures, given by international authorities, will range on
scientific topics from mathematics and physics through to biology
and economics. The books are intended for a broad audience of
graduate students and faculty members, and are meant to provide a
'mise au point' for the subject they deal with.
The symbol of the Accademia, the Lynx, is noted for its sharp
sightedness; the volumes in the series will be penetrating studies of
scientific topics of contemporary interest.

Already published

Chaotic Evolution and Strange Attractors: D. Ruelle
Introduction to Polymer Dynamics: P. de Gennes
The Geometry and Physics of Knots: M. Atiyah
Attractors for Semigroups and Evolution Equations: O. Ladyzhenskaya
Half a Century of Free Radical Chemistry: D.H.R. Barton

Bound carbohydrates in nature

LEONARD WARREN

The Wistar Institute Philadelphia, USA

CAMBRIDGE
UNIVERSITY PRESS

CAMBRIDGE UNIVERSITY PRESS
Cambridge, New York, Melbourne, Madrid, Cape Town,
Singapore, São Paulo, Delhi, Mexico City

Cambridge University Press
The Edinburgh Building, Cambridge CB2 8RU, UK

Published in the United States of America by Cambridge University Press, New York

www.cambridge.org
Information on this title: www.cambridge.org/9780521442312

© Cambridge University Press 1994

First published 1994

A catalogue record for this publication is available from the British Library

ISBN 978-0-521-44231-2 Hardback
ISBN 978-0-521-44743-0 Paperback

For Eve

Contents

Preface

This short book does not pretend to be a scrupulously documented, comprehensive survey of our vast and expanding knowledge of the bound carbohydrates; it is neither a broad survey nor is it detailed. The reader is often referred to a review that covers a topic under discussion so that full credits are not given in detail. I have covered only those topics with which I have grappled trying to acquire new information by experimentation and analysis. I include other subjects because they have, over the years, intrigued me. For the most part, what is presented here is the subject matter of six lectures presented to the Italian Academy of Sciences at the University of Siena, Italy, 27–30 April 1992. I am particularly grateful to Professor Baccio Baccetti of the University of Siena for his kind invitation and to IBM and the University for their support.

What I present closely reflects and delineates my career in biochemical research at the National Institutes of Health, Bethesda, MD and then at the University of Pennsylvania and the Wistar Institute, Philadelphia, PA. The pattern of thinking and the experimental approach of a research worker is often strongly influenced by a first experience in research: an imprinting process. So it was with me when I began to work in the 1950s in the laboratory of J.M. Buchanan at MIT establishing some reactions involved in purine biosynthesis. Those were exciting days, immersed as I was in the lore of enzyme reactions and intermediary metabolism, and from this source evolved my subsequent interests and biases. I hope that this small book captures some of the flavor of my daily concerns.

LEONARD WARREN
Philadelphia, April 1992

Acknowledgements

I am indebted to my long-standing colleagues H. Tabor, S.S. Cohen, C.A. Buck, M.C. Glick, and J.-C. Jardillier, and to the many students and technicians who either knew of my work in fact, or helped carry out the research described in these lectures.

1

The carbohydrates of glycoproteins

Living matter consists largely of water and complex polymers of amino acids, lipids, nucleotides and carbohydrates. Carbohydrate polymers are special in that they are usually associated with the three other polymers. They are stably linked with amino acid polymers (proteins) or with lipids as glycolipids that participate in non-covalently-linked associations, the lipid bilayer of biological membranes. DNA and RNA are, in essence, polymers of D-ribose-phosphate and 2-deoxy-D-ribose-phosphate to which are attached purine and pyrimidine bases at the C-1 reducing position. While the synthesis of proteins and nucleic acids is guided by a template to maximize fidelity of product formation, synthesis of carbohydrate and lipid membrane polymers, which perform a more structural function, is not. As will be discussed, a variability and relative lack of quality control in the making of lipid and carbohydrate polymers may have its special role in the operation of organisms in an environment that is variable and unpredictable.

Sugars almost certainly existed before life itself appeared on Earth. It has been known for over a century that many sugars can be formed from formaldehyde in alkaline solution. Condensation of formaldehyde, most probably a prebiotic constituent, has been carried out in the laboratory to form glycoaldehyde, trioses, tetroses, pentoses, and hexoses. Polymerization of formaldehyde can be catalyzed by surfaces provided by insoluble silicates and carbonates. UV light, electric discharge and ionizing radiation at the right pH and temperature may also promote synthesis of sugars. The phosphorylation of sugars is another indispensable step in the assembly of the building blocks of living matter. Cyanogen, a likely prebiotic constituent, is capable of catalyzing the synthesis of glucose mono- and diphosphate from glucose and orthophosphate.

Several alternative, synthetic pathways and reaction conditions have been proposed but specific details will probably never be established. However, it is highly probable that there was a relative abundance of various sugars and their phosphates in the prebiotic world. In this soup, the basic building blocks must have polymerized and assembled, ultimately to form a self-sustaining, self-reproducing, adaptive entity. Our credulity is strained in accepting that such an awesome sequence took place, but, over inconceivably great periods of time, it did occur. The possibility that carbohydrates will be found in fossils is almost zero because they are relatively unstable, capable of being dehydrated and of combining with other molecules. At higher temperatures they caramelize and char.

The structural diversity possible by linking the different, common sugars is immense: theoretically far greater than that of proteins which largely consist of 22 amino acids linked by a single type of union the peptide bond. Linkages between sugars can occur through a glycosidic linkage between the anomeric, first carbon of a sugar in either an α or β configuration with any of a variety of hydroxyl groups on the adjacent sugar. The number of possible forms that can be produced from two or three of the same sugars or amino acids is shown in Table 1. In fact, many possible combinations of sugars do not seem to exist.

Recurring structural patterns are found that are probably dictated by the limited specificities of sugar transferases and the stringent prohibitions of structural requirements. While polypeptides are linear and unbranched, sugar polymers may branch in many ways, so that the total number of surface conformations of saccharides and the information they bear can be enormous. In general, the components of sugar polymers are fixed with little or no rotation about glycosidic linkages. Because of these properties, saccharides bound to protein and lipid are often immunological epitopes and are the target for combination with lectins.

Table 1. *Diversity of dimeric and trimeric forms of saccharides and peptides*

Composition	Number of isomers	
	Saccharides	Peptides
a-a (dimer)	11	1
a-a-a (trimer)	176	1
a-b-c (trimer)	1056	6

Note: From Clamp, 1974.

They are involved in receptor activities for hormones, viruses, bacteria, and toxins.

History[1]

Human beings have always been aware of sticky, slimy and viscous substances in animal and vegetable matter. They were described in Latin works and were seriously investigated and discussed by 18th century scholars. As early as 1747, Beccaria, an Italian priest, found a gluten in wheat which was shown by Taddei in 1819 to consist of several components, one of which was a 'schliemige Materie'. Workers in the field distinguished this material from others that were essentially protein. This important distinction was reinforced by German workers who found that mucus, but not protein, could be precipitated at low temperature by acetic acid. In 1805, John Bostock published the results of systematic experiments on the nature of mucins, as well as albumen and gelatin.

In 1865, Eichwald discovered that sugar was a component of mucin. Mucins from ovarian cysts and a wide variety of tissue, when hydrolyzed with mineral acid, released a reducing substance that was considered to be glucose. In the next 25 years, a clearer

[1] Much of this section is based on material in Gottschalk, 1960, 1972.

view of the nature of mucins emerged from the work of Hoppe-Seyler, Giacosa and others, as it became apparent that there was more than one sugar (glucose) present. Some contained nitrogen and had acidic properties that were probably due to sialic acid. It was also shown that the sugar and protein components were stably linked by a covalent bond. With the development of new, analytic methods for carbohydrates and the evolution of organic chemistry, the structures of hexoses, hexosamines, uronic acids and, eventually, sialic acids were established.

However, progress was slow in this difficult and obscure field because too much groundwork had to be done. The general lack of interest in the area was certainly recognized by one of the outstanding workers in the field, the late Karl Meyer, when he stated 'I work on the gunk and muck that others throw away'. Meyer, and his colleagues before him, worked on complex, multicomponent, unknown, heterogeneous materials, using methods that were poorly understood; this frequently led to false interpretation and error, even by the greatest of researchers. Despite the importance and beauty of carbohydrate biochemistry, the image of groups of complicated sugar molecules with hydroxyl and amino groups sticking out in every direction, has discouraged interest and has even frightened people away from the field.

The nature of both the carbohydrate and protein polymers and their linkages had to be clarified. It was only in 1930 that Sorensen clearly established that the carbohydrate component of the glycoprotein ovalbumin was heterogeneous. By fractional crystallization of highly purified ovalbumin, he obtained more soluble preparations that contained 50 times more carbohydrate per mole than early-crystallizing, less soluble fractions. Heterogeneity of pure carbohydrate polymers was always a source of confusion.

Brilliant studies on the dynamic nature of cell constituents and the elucidation of metabolic pathways (glycolysis, Kreb's cycle, the synthesis of amino acids, purine and pyrimidines, lipids, etc.) were of direct relevance to the glycoprotein field. The identification and synthesis of activated sugars by the Leloir group triggered an explosion of investigations of general interest on the biosynthesis and interconversions of sugars and on the step-by-step synthesis of sugar polymers.

Although structural work was not of broad concern to the scientific community, there were two lines of investigation that drew attention to protein-bound carbohydrates. Work on blood group substances, an amalgam of glycoprotein chemistry with immunology and genetics, was widely followed by the scientific community. While blood group substances were discovered by Landsteiner at the turn of the century, 50 years passed before their carbohydrate specificities were elucidated by Morgan and Watkins. The complete sequences of the enzymes responsible for specificity have been revealed only in the last decade of the 20th century.

There was general interest as well in the study of influenza virus interacting with red blood cells which led to the discovery of neuraminidase (sialidase) and the characterization of the sialic acids. More recently, the surface membrane of cells, rich in carbohydrate-containing glycoprotein that undergoes characteristic changes in malignancy and other diseases, has been studied intensively. In recent years, major advances have been made in our understanding of intercellular adhesiveness and the molecules participating in the process.

Another major chapter is that concerning proteoglycan (mucopolysaccharide) biochemistry, with work on the structure of chitin from which glucosamine is derived. This hexosamine was also found in ovarian cysts, ovomucoid and seromucoid from horse serum. Galactosamine and D-glucuronic acid were shown to be major components of chondroitin sulfuric acid from cartilage. Over many years, heparin, heparan sulfate, hyaluronic acid and chondroitin A, B and C, keratan sulfate, dermatan sulfate, and other heteropolysaccharides have been identified and chemically characterized. A good understanding of their synthesis and function has been achieved.

Our present knowledge of mucins and glycoprotein biochemistry had to await the development of fractionation and purification methods, sensitive analytic procedures for amino acids and sugars, and methods for the determination of molecular weights of macromolecules. Major advances have been made with the advent of amino acid sequencing techniques, isotopic methods, gas and high-performance liquid chromatography, electrophoretic methods, fast atom bombardment and high-resolution NMR spectroscopy, and the use of both exo- and endoglycosidases. The molecular biological revolution has had a growing, enormous impact on glycobiology both in its structural and functional aspects, providing information and insights that were undreamt of even two decades ago.

One of the liveliest areas of research, at the present time, is on glycoproteins in and on the surface membrane of the cell which function as receptors for hormones and drugs and as components of the adhesive process. The structure of the molecules and their function have been delineated by immunocytological and powerful molecular biological methods such as site-directed mutagenesis.

Microheterogeneity of bound carbohydrates

Most membrane and extracellular proteins bear sugar groups. However, some notable exceptions are serum albumin, trypsin and chymotrypsin. Analysis of purified glycoproteins reveals that the variety of individual molecular species with a common polypeptide backbone can be enormous. The groups may be bound to the hydroxyl group of serine, threonine or hydroxyproline by a glycosidic bond or to the amide group of asparagine by a glycosylamine linkage. There are a few, rare instances of S-glycosidic bonds such as in digalactosyl cysteine found in human urine or triglucosyl cysteine in human red blood cells. Other relatively uncommon linkages are L-fucose linked by an α-glycosidic bond to the hydroxyl group of serine or threonine and mannose linked to these amino acids in yeast and fungi. Direct linkage of D-glucose to serine and threonine, characteristic of animal cell nuclei, has also been found (Hart *et al.*, 1989). The same amino acids may not be glycosylated in members of the same glycoprotein species. The number of groups may vary from one, as in ribonuclease and ovalbumin, to hundreds, as in mucins, and the groups themselves may consist of two to 30 sugars or may, in fact, be large glycosaminoglycans (Baker *et al.*, 1980). In many instances, groups of sugars may be bound to both serine and asparagine in the same polypeptide, as in glycophorin. Here there are 15 groups of the former and one of the latter (Tomita and Marchesi, 1975).

Groups attached to a specific amino acid may differ in structure. Exhaustive digestion of a purified glycoprotein by proteolytic enzymes may produce far more glycopeptide species than there are binding sites on the polypeptide. Quantitative analysis of the sugars comprising a glycopeptide almost invariably results in ratios of sugar that are not integral, which can only mean that the glycopeptides are heterogeneous in their carbohydrate component.

Although porcine ribonuclease has only one glycosylation site, eight forms of the molecule can be found in pancreatic secretion differing in their oligosaccharide. One form of ribonuclease is devoid of carbohydrate while one other contains sialic acid (Plummer and Hirs, 1964; Beintema *et al.*, 1976). Although ovalbumin also has only one glycosylation site, an array of glycopeptides can be derived from purified ovalbumin from one egg (Lush, 1961). The same groups are found in various proportions in ovalbumin from ostriches, gulls, turkeys, and ducks (Lush and Conchie, 1966). A third example of heterogeneity of carbohydrate group can be found in the surface membrane glycoproteins of the baby hamster kidney cell BHK21/C-13 (Baker *et al.*, 1980). At least 12 carbohydrate groups can be identified from a single glycoprotein. If all were present on each polypeptide, their combined molecular weights would be almost that of the entire glycoprotein instead of a known 10% of the molecular weight. Obviously, not all carbohydrate groups are present on each polypeptide. These are but a few of many examples of microheterogeneity.

Microheterogeneity does not result from random degradation of glycoproteins during their isolation. In many studies, every precaution was deliberately taken to avoid degradation. During synthesis carbohydrate polymers are assembled in the absence of a guiding template. Fidelity is dependent upon enzyme specificity and this may not be absolute. Since conditions such as cation and anion concentrations, intracellular pH, availability of substrate, and other variable factors can affect enzyme activity, the final products at the end of a multienzyme process may not be identical. The relative amounts of competing enzymes can determine the pathway to be followed. Greater variation may occur toward the non-reducing ends of the chains, which often involves L-fucose and sialic acids which are 'cappers' or chain terminators.

However troubling heterogeneity is to the analyst, it appears to be a fundamental characteristic of carbohydrate polymers. The cell usually shows a remarkable tolerance to variations in the carbohydrate component of glycoproteins. The proposal will be made that the synthesis of carbohydrate groups, responding to environmental conditions in their broadest sense, may be the basis of a plastic, adaptive, survival process operating through a shifting polymorphism.

Function of bound carbohydrates

In recent years, our knowledge of the structure and function of the bound carbohydrates has been increasing at an alarming rate. Large numbers of protein-bound carbohydrates have been described. Their roles in nature are complex and cannot be simply categorized and, frequently, experimentalists are unable to assign any function to this component. In some instances, their function is associated with the general properties inherent in their molecular structure: for instance, they constitute a hydrophilic mass that may be charged with sulfate, phosphate or sialic acid residues that readily form hydrogen bonds and increase the solubility of proteins. In globular proteins, they protrude into the aqueous phase at the surface of the polypeptide that is folded in on itself and is stabilized in the folded form by hydrophobic interactions between interior amino acids. This latter interaction is the major factor responsible for folding. Depending on their size and nature, the external carbohydrate groups could modify, promote or deter folding. Programed glycosylation of nascent polypeptides within the Golgi could have some effect on folding, resulting in a desired conformation. Specific conformations may prevent certain amino acids from being further glycosylated, or render others available for glycosylation. Variation in polypeptide folding induced by carbohydrate groups could possibly be the basis for small shifts of functional activities (see p. 20). The carbohydrate groups may be in essence multiple, covalently bound, allosteric effectors. Each site on the polypeptide may bear no group, or one of a variety of carbohydrate groups in accordance with known microheterogeneity. Here, they could be modifiers of activity rather than critical effectors.

In a mucin such as that of ovine submaxillary gland, hundreds of carbohydrate groups exist largely as dimers along the polypeptide. They are charged and hydrophilic, preventing the folding due to hydrophobic association of amino acids within the chain. Sialic acids, linked to *N*-acetylgalactosamine residues along the polypeptide, interact with the amino acids to stiffen the chain. The fixed charges of sialic acids repel each other to stretch the polypeptide, creating rigid, linear molecules that get tangled up with each other, especially at low ionic strength. This is the basis of the viscous, lubricating, and protecting properties characteristic of mucins. When mucins in solution are treated with sialidase, charges are removed leading to a collapse of the extended molecule and a rapid fall in viscosity (Gottschalk and Thomas, 1961). Surface membrane glycoproteins may bear carbohydrate groups on the outside of the cell to stiffen the polypeptide emerging from the lipid membrane so that some active, functional groups may be held away from the tangle of molecules at the immediate cell surface.

Carbohydrate groups, in some instances, are very large and bulky. Since they can be great space occupiers, they are common constituents of intercellular matrices and basement membranes. If they bear charged groups in these locations, they may also function as ion exchangers or adsorbents, regulating and stabilizing the composition of intercellular fluid. Sialic acids which are usually highly exposed, can bind cations such as Ca^{2+}. Surprisingly, the glycerol side chain (C7–9), as well as the carboxyl group, is intimately involved in the binding (Jaques *et al.*, 1977).

In the 1980s, the elements of the extracellular matrix have rapidly become known, with new insights into the mechanism of adhesiveness during proliferation and differentiation. It is apparent that there are important, specific, ionic interactions between carbohydrate-rich glycoproteins, such as fibronectin and cadherins, with sulfated and other glycosylaminoglycans, growth factor receptors, and growth factors such as fibroblast growth factor. Some fragments of sulfated glycosylaminoglycans enter the nucleus where they may affect DNA transcription.

The presence of carbohydrate groups along a polypeptide can change its function (Table 2). They may protect it from degradation by proteases, not only because the carbohydrate may induce folding that buries more susceptible regions of the chain but also because it masks the chain. It was observed soon after neuraminidase was discovered that certain glycoproteins rich in sialic acid could be degraded by trypsin, only after sialic acid was removed by sialidase. In a sense, the sialic acid and underlying carbohydrate residues form a charged molecular shield or skin. Since bound carbohydrates can prevent degradation, it is possible that they may be regulators of the rate of protein turnover. The removal of certain sugars by glycosidases, or the programed cessation of glycosylation within the cell or organism, may be a key, permitting a timed and orderly degradation of proteins, which can take place during the mitotic cycle.

The bulk of protein-bound carbohydrate is at the surface of the cell. In the L cell, a mouse fibroblast, at least two thirds of the glycolipid sialic acid is in the plasma membrane. When appropriately stained, a layer of carbohydrate-containing material can be seen extending as much as 15 nm out from the lipid bilayer of some

Table 2. *General functions of carbohydrate polymers*

Viscosity
Lubrication
Protective coating
Protection against proteolysis
Increase solubility in water, decreased precipitability
Antifreeze
Structural function (matrix, basement membrane)
Energy storage (starch, glycogen)
Provide bulk (matrix, basement membrane)

cells. This highly hydrated tangle may have a protective function and may limit diffusion of molecules into cells or along the surface of the cell. The masking and stiffening function of carbohydrate groups on protein at the surface of the cell may be responsible for the prevention of an interaction between one cell and another, or between the cell and soluble macromolecules such as antibodies. In general, there is more interfering glycoprotein on the surface of malignant cells, which could be responsible for their partial or complete escape from immune surveillance and their decreased susceptibility to natural killer cell-mediated cytotoxicity.

Because most extracellular proteins bear carbohydrate groups, it was postulated that bound carbohydrates are a signal for secretion or are part of the secretory mechanism (Eylar, 1965). However, albumin, trypsin, and chymotrypsin are secreted and are free of sugar while, conversely, there are many intracellular glycoproteins.

In addition to these non-specific but important activities of protein-bound carbohydrates, there are several processes known where the specific structure of the carbohydrate involved is critically important. Some examples of this specificity will be discussed, but it should be kept in mind that there are really no underlying principles or general rules for predicting where and what specific sugar structures will be required in a particular process. Further, whatever the general or particular role of bound carbohydrates may be, the activity may change with time in a programed manner during the cell cycle or during development.

Although the list of examples of acute and specific requirements for glycoprotein sugars is growing, there are countless studies, probably mostly unreported, where tampering with sugars, by treatment with glycosidases or pharmacological agents, has no apparent effect or only 'minor' effects. The mere presence of a carbohydrate group, regardless of its size and shape, may be all that is required. The cleavage of 80% of the sugar from interferon molecules does not alter their antiviral activity or their antibody-binding properties (Bose *et al.*, 1976), nor is there a change in enzyme activity of penicillium nuclease P1 and mung bean nuclease I after removal of saccharide by endo-β–acetyl-glucosaminidase H (Trimble and Maley, 1977). These common, disappointing findings have given rise to the feeling that the bound sugars should 'do more'. One can postulate that functional changes are induced by altering carbohydrates and, though small, these cumulatively are very significant, as for example in the removal of carbohydrates from thyroxine-binding globulin (Cheng, Morrone and Robbins, 1979). When this serum glycoprotein is purified and treated with sialidase, β-galactosidase, α-mannosidase, and β-*N*-acetylglucosaminidase to remove 80% of its sugars, the product binds as much L-thyroxine as the untreated glycoprotein. However, the affinity constant is reduced from 1.6×10^9 M^{-1} to 0.58×10^9 M^{-1}, a relatively minor change, but within the cell even smaller changes may have important consequences. Although the removal of bound carbohydrates may have little or no effect on a function that is being tested, e.g. an enzyme activity, the carbohydrate may be involved in other processes such as homing, protection of a molecule against degradation, secretion or interaction with other macromolecules. 'Small' changes of 5–20% are usually ignored by experimentalists, especially if methods of assay are not completely reproducible and the assay conditions are not optimal for revealing functional changes.

In our own studies (Rothman *et al.*, 1989), monoclonal IgG

against a cell surface antigen of tumor cells was synthesized by hybridomas exposed to castanospermine, an inhibitor of glycosylation and carbohydrate processing. This altered IgG and the unaltered IgG bound equally well to the cell surface antigen. However, lymphocyte-mediated antibody-dependent cytotoxicity was enhanced significantly when target cells were sensitized with IgG bearing the atypical carbohydrate rather than the native IgG; altered fucosylation on the core *N*-acetylglucosamine residue of IgG may be responsible for the enhanced cytotoxicity.

Inducible changes in bound carbohydrates

As previously stated, the biosynthesis of carbohydrate polymers is not guided by a template and so there may be considerable variation in final products, depending on the availability of substrates and cofactors and conditions within the cell. The relative amounts and activities of many enzymes, both synthetic and degradative reflecting the genetic properties of the cell and the control of gene expression, may be determining factors. Yet, whatever the final mix of sugar polymers on polypeptides, the cell survives and prospers. In contrast, single amino acid substitutions in a polypeptide frequently have serious consequences and can be lethal.

The heterogeneity of the saccharides of serum glycoproteins are well known. Within the organism, heterogeneity may be a natural consequence of the different states of the body during the course of the day. Glycoproteins are made by liver cells located in various parts of the organ that may have different blood supplies and each group of cells may have a different pH and state of oxygenation. Exposure to materials from the gut may vary. Liver cells may be synthesizing serum glycoproteins during fasting periods or when engorged with carbohydrates and lipids, during sleep or exercising, and in each state the bound carbohydrates made could be slightly different. All would be pooled in the blood to constitute populations of glycoproteins with heterogeneous, bound carbohydrates.

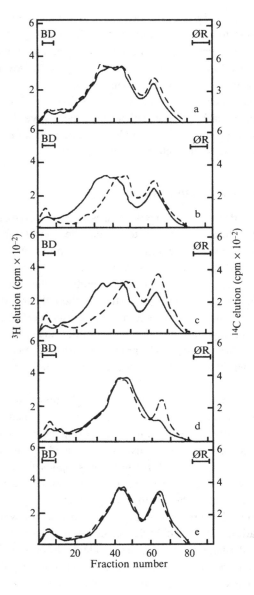

The carbohydrate composition of the coat glycoprotein of vesicular stomatitis virus depends on the cell in which the virus is produced. Virus was grown in several lines of mammalian cells in culture (mouse, hamster and dog). Although both the proportion of sugar in the viral glycoprotein and its size were approximately the same, its carbohydrate composition differed (Etchison and Holland, 1974). Similarly, the sialic acid contents of Sinbis virus grown in hamster and chick cells are different (Strauss, Burge and Darnell, 1970), and when the virus is grown in insect cells sialic acid is absent because sialic acid is not made by insects. Since the virus depends on the glycosylation mechanism of the host cell, it is possible that viral tropism can vary with the species and the tissue specificity of the cell in which the virus is produced.

The bound carbohydrates of cells in culture can be induced to change by altering their environment. For instance, there are dramatic differences in the bound carbohydrates of chick embryo chondroblasts grown in suspension culture, in monolayer, and as a pellet culture (Cossu *et al.*, 1982a) (Fig. 1).

Comparisons of the protein-bound carbohydrates of chondroblasts were made by the double-label method, in which one of the

Fig. 1. Changes in the bound carbohydrate of glycoproteins in cells grown in different conditions. Double-labeled elution patterns of [^{14}C] and [^{3}H]fucose-labeled glycopeptides derived from glycoprotein of chick embryo chondroblasts grown under different conditions. Glycopeptides were eluted from a column of Sephadex G50 (1 × 80 cm) with 0.01 M ammonium acetate in 20% ethanol. Markers: BD, blue dextran; øR, phenol red. (*a*) Monolayer culture of chondroblasts (——); suspension culture of chondroblasts (- - -). (*b*) Monolayer culture of chondroblasts (——); cartilage organ culture (- - -). (*c*) Monolayer culture of chondroblasts (——); chondroblast pellet culture (- - -). (*d*) Cartilage *in vivo* (——); cartilage organ culture (- - -). (*e*) Cartilage *in vivo* (——); chondroblast pellet culture (- - -). (Taken from Cossu *et al.* (1982). *J. Biol. Chem.* **257**, 4463, with permission.)

cell cultures is grown in the presence of a [^{14}C] sugar (e.g. L-fucose or D-glucosamine) and the other culture is grown with the same sugar labeled with ^{3}H. After incorporation over 15 h, cells containing the same amount of incorporated radioactivity are mixed and homogenized. The homogenate, or isolated glycoproteins from the homogenate, are exhaustively digested with pronase that destroys the polypeptide but leaves radioactive glycopeptides. These are separated on a long column of Sephadex G50 (or P10). Both ^{14}C and ^{3}H are determined in each tube so that each is the other's control. The data are processed and plotted by computer. It is apparent from the data of Fig. 1 that the glycopeptides derived from chondroblasts rapidly growing in suspension or in monolayer are considerably larger than those from chondroblasts *in situ*, in organ culture, or in a compact pellet. The larger glycopeptides elute earlier (*a*, *b*, *c*).

Similarly, when hamster cells are grown in culture on glass, plastic or plastic coated with collagen, the populations of bound carbohydrate are clearly different (Warren, Blithe and Cossu, 1982). The differences could be explained by differential productions of specific glycoproteins, each with its own complement of carbohydrate groups. However, it was found that these carbohydrate groups were not the same, even in individual glycoproteins from cells grown under different conditions.

The buffer in which cells are growing in culture may influence the carbohydrate composition of glycoproteins. When murine carcinoma cells are grown in different media with 1 or 10% fetal bovine serum, or in chemically defined medium with or without Hepes buffer, the bound carbohydrates of the cells differ. Further, the composition in 1–15 day cultures is not the same as that of 16–39-day-old cultures (Megaw and Johnson, 1979). It has been observed that the bound carbohydrates of hamster cells differ slightly when the cells are growing at pH 6.5 and 7.2 (Warren, unpublished). By the double-label method, differences have also been observed in the bound carbohydrates of the fathead minnow, grown in culture at 23 and 33 °C. This cell has a broad temperature tolerance (Warren, unpublished).

The level of sialic acid in the epithelial tissue cells of the Prussian carp (*Carassius auratus*) and other fish is higher when the fish are kept in freshwater as compared to seawater but the higher level could be due to enhanced synthesis of sialic acid-rich glycoproteins rather than increased amounts of sialic acid per polypeptide (Hentschel and Muller, 1979). Finally, glycoprotein carbohydrates of cells in culture may be altered by agents such as ethidium bromide, 2-deoxyglucose, cyclic cAMP, sodium butyrate, tunicamycin, and other drugs, as well as in retinoic acid deficiency (Warren *et al.*, 1983). It is evident that environmental conditions and many factors and reagents can change the saccharide component of polypeptides.

Programed changes in bound carbohydrates

Differences in populations of bound carbohydrates may depend on the state of growth of the cell (Buck, Glick and Warren, 1971; Muramatsu *et al.*, 1973). During growth, certain populations of bound carbohydrates increase in size. The same shift occurs and, more importantly, persists in the glycoproteins of malignant cells. If this change in bound carbohydrates closely reflects some aspect of cell division, it would suggest that the malignant cell is in a permanent growth state. Changes in the bound carbohydrates in malignancy will be discussed later (p. 62). Marked differences in the glycolipids of growing and resting cells have also been observed (Hirschberg, Wolf and Robbins, 1974).

Some carbohydrate groups of glycoproteins are developmentally regulated so that they change in structure and function in an orderly manner during embryogenesis. The surface glycoprotein and fibronectin from adult 3T3 cells and differentiated embryonal carcinoma cells each have M_r values of 220 000. However, the parental mouse embryonal carcinoma cells F9 produce a fibronectin of M_r 250 000. The 30 000 M_r difference is due to an extra chain of covalently-bound heparin sulfate (M_r 20 000) and one of lactosylaminoglycan (M_r 5000–10 000) on the fibronectin in addition to the usual complex, biantennary carbohydrate structures (M_r, 2000–3000) whose synthesis is somewhat increased (Cossu and Warren, 1983). Embryonic cells, such as F9, adhere poorly to each other and to a plastic surface. It is possible that the long and negatively charged carbohydrate chains on fibronectin located at the cell surface interfere with the adhesive process, such as between fibronectin and the glycosylaminoglycan that occurs in basement membrane and in intercellular matrix of adult tissue. When synthesis of these long molecules on fibronectin ceases, a programed adhesiveness may commence. Polylactosamine chains on human fetal placental fibronectin are known to interfere with binding to gelatin and they also protect the molecule against proteolytic digestion (Zhu and Laine, 1985). However, in another study, the long chains of fibronectin appeared to increase the affinity of the molecule to gelatin and heparin in the presence of divalent cations.

A programed change in the carbohydrate component of the neural cell adhesion molecule (N-CAM) also takes place during development. This molecule is characterized by an almost unique, bound polysialic acid component. It is present in relatively large amounts in embryonic brain and developing kidney and is sharply reduced in the adult tissue (Finne, 1985). Reduction is accompanied by an increased adhesiveness and, in nerve tissue, may produce changes in the levels of neurotransmitter enzymes, in the functional associations of the axons, and in intercellular communications (Rutishauser *et al.*, 1988). A third example of programed, differential glycosylation occurs when the human promyelocytic leukemic cell HL60 is induced to differentiate into a macrophage by phorbol ester. Large carbohydrate groups on glycoproteins of the cell surface, such as polylactosamine, which are partially degradable by endo-galactosidase, are made in reduced amounts; within hours, the cells become adherent. On the other hand, when HL60 cells are induced to differentiate into granulocytes by exposure to dimethyl sulfoxide, hypoxanthine or retinoic acid, the bound carbohydrate groups are unaltered and the cells do not become adherent (Cossu *et al.*, 1982b). Programed changes also occur in the gangliosides of brain, in intestinal mucosa and in other tissues during development.

A possible role for bound carbohydrates

A large and complex array of carbohydrate oligomers are bound in glycoproteins of vertebrates. Their staggering diversity makes it difficult, if not impossible, to generalize as to their function and their role in nature. Yet they have displayed a rather stable pattern of distribution over vast periods of time and in many species (fish, mice, humans), which suggests that their presence is critically important. Frequently, however, the experimentalist finds that alterations of carbohydrate groups by glycosidases or by pharmacological agents have little or no effect on activity. Paradoxically, though indispensable and present in all forms of life, the cell can tolerate a surprising degree of variation in its structure, since essentially all glycoproteins are, as discussed earlier, heterogeneous in their carbohydrates and these groups can change with shifting environmental conditions.

The behavior of glycolipids is a striking example of this tolerance. There is an enormous diversity of glycolipids, several hundreds in animal cells. These have carbohydrate and lipid structures that, in some instances, change in a very specific, programed manner during embryonic development. They are carriers of blood group and tumor antigens and are receptors or can modulate receptor activity at the cell surface. They are probably an important determinant of the character of the membrane itself (charge, permeability).

When cells or whole fish embryos (Medaka) are grown in the presence of PDMP (1-phenyl-2-decanoylamino-3-morpholino-1-propanol), a structural analog of ceramide and an inhibitor of glycosylceramide synthetase, glycolipid synthesis is sharply curtailed, metabolic labeling with radioactive sugars is reduced by over 75%, and the glycolipid content and several tissue-specific glycolipid antigens are dramatically reduced. Remarkably, the embryos remain viable and develop without obvious abnormality for at least 10 days (Fenderson *et al.*, 1992). In this instance, the tolerance for quantitative and probably qualitative change is extreme. It is apparent that the carbohydrates (and lipids) are not involved in acute functions, despite the specificity of structure and programed change in structure they may undergo. Rather, as will be discussed, the carbohydrates of glycolipids, in their variety, quantity and pervasiveness, must serve a low key background function, perhaps as a survival factor during evolution.

The following hypothesis concerning glycoproteins is advanced to resolve this apparent contradiction. Our speculation also applies to the modifying functions of glycolipids. Whatever the activity of a glycoprotein, whether it is an enzyme, hormone, structural agent, etc., the homogeneous polypeptide moiety which is responsible for the activity is slightly modified by the particular combination of oligosaccharides it happens to bear. Structural differences result in functional differences, however small, and, since the carbohydrates display heterogeneity, the glycoprotein exhibits a range of activities and responses to conditions. I suggest that whatever the environment in which the glycoprotein agent acts, there are always some polypeptides bearing combinations of saccharides that operate more advantageously, whether the varying environmental factor is pH, temperature, concentration and state of the substrate, cations and anions, or other macromolecules. This constitutes a preadaptive advantage for the organism that improves the chance for survival. A specific array of carbohydrate groups on a polypeptide could be selected for during evolution. The difference in activity of polypeptides bearing slightly different carbohydrate groups may be small and, from an experimentalist's point of view, might be trivial. However, within an evolutionary time period, the small differences and survival advantages could prove to be important. Further, with a new environmental condition, there may be a different mix of bound carbohydrates, a shifting polymorphism that works to the advantage of the organism.

Changes may occur during the development of an organism in a genetically programed manner. This also holds for other post-translational modifications such as phosphorylation, acetylation or sulfation. The response of glycolipid to the environment may be particularly lively and of special, but unknown, significance because both its lipid (fatty acids) and carbohydrate components can change with conditions. In a sense the carbohydrate groups track the environment, resulting in a more protective array of responses. However, there may be an absolute need for some specific carbohydrate structures with no tolerance for deviation. In this instance, carbohydrates, by chance, may have become a critical part of a mechanism. Where this has occurred is completely unpredictable.

What are the corollaries of such a hypothesis? If bound carbohydrates function as an environmental buffer, then they should be found where the cell has least control over the environment, i.e. at the cell surface. 60–70% of the cell glycoproteins are exposed at the cell surface, on the outside of the lipid bilayer of the plasma membrane. Within the cell there should be more control and here there is much less bound sugar. Most intracellular enzymes are sugar-free proteins. In this regard, it is probably significant that enzymes devoid of sugar are involved in the citric acid and glycolytic pathways, and in the synthesis of purines, pyrimidines, amino acids, DNA or RNA. Notable exceptions are the lysosomal enzymes which act on internalized bits of the extracellular world or on extracellular material after secretion. Within the membrane-bound nucleus, control is most rigorous and there is relatively little bound sugar, while in the strictly extracellular world, there is an abundance of carbohydrate-rich, glycosylaminoglycans and glycoproteins.

Small and non-acute changes in bound carbohydrate occur and may lead to non-lethal changes in function. Thus, chronic shifts in cellular environment during aging and in pathological processes, such as atherosclerosis, degenerative diseases, diabetes and other metabolic diseases, may lead to shifts in structure and function of glycoproteins. The changes may be silent and unrecognized but ultimately, with time, detrimental activity may be cumulative, breakdown can occur and disease become manifest. Similar events could take place in acute disease. The cells in and around a focus of inflammation are exposed to new conditions of temperature, pH, and arrays of metabolites, which may result in the synthesis of altered glycoproteins. As I will discuss later, it has been found that when cells become malignant there is a chronic enlargement of the carbohydrate groups of 80% of the glycoproteins associated with every membrane system of the cell (p. 62).

2

Derivatives of neuraminic acid

Sialic acids

The sialic acid *N*-acetylneuraminic acid was first isolated and crystallized by Blix (1936). Since it came from salivary gland mucin, he named it 'sialic acid'. Klenk (1941) found it in human brain and called it neuraminic acid, while Yamakawa and Suzuki (1952) discovered it in red blood cells and called this compound 'hemataminic acid'. The identity of these materials, derived from different sources, was ultimately recognized by detailed comparison of their properties. By convention, the term 'sialic acid' is generic, comprising a family of compounds which have in common neuraminic acid (Neu) (5-amino-3,5-dideoxy-D-glycero-D-galactononulosonic acid). This 9-carbon, carboxylic acid does not exist in nature, but many of its derivatives have been described. The open-chain keto form, and the naturally occurring, closed pyranose ring form of *N*-acetylneuraminic acid (Neu5Ac) are depicted below in three standard ways (Fischer, Haworth and Reeves) (Scheme 1).

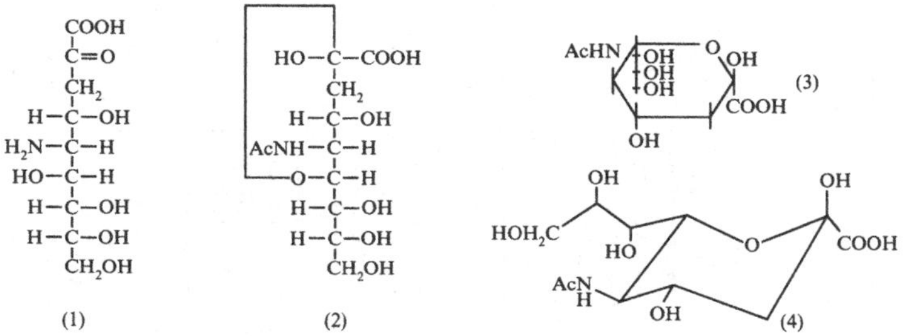

Scheme 1. Neuraminic acid in the open, keto form (1) and three depictions of the pyranose ring structure of *N*-acetylneuraminic acid: Fischer (2), Haworth (3), and Reeves (4).

Table 3. *Forms of sialic acid*

Name	Abbreviation	Substituent				
		R^4	R^5	R^7	R^8	R^9
N-Acetylneuraminic acid	Neu5Ac	H	Acetyl	H	H	H
N-Acetyl-4-*O*-acetylneuraminic acid	Neu4,5Ac$_2$	Acetyl	Acetyl	H	H	H
N-Acetyl-7-*O*-acetylneuraminic acid	Neu5,7Ac$_2$	H	Acetyl	Acetyl	H	H
N-Acetyl-8-*O*-acetylneuraminic acid	Neu5,8Ac$_2$	H	Acetyl	H	Acetyl	H
N-Acetyl-9-*O*-acetylneuraminic acid	Neu5,9Ac$_2$	H	Acetyl	H	H	Acetyl
N-Acetyl-4,9-di-*O*-acetylneuraminic acid	Neu4,5,9Ac$_3$	Acetyl	Acetyl	H	H	Acetyl
N-Acetyl-7,9-di-*O*-acetylneuraminic acid	Neu5,7,9Ac$_3$	H	Acetyl	Acetyl	H	Acetyl
N-Acetyl-8,9-di-*O*-acetylneuraminic acid	Neu5,8,9Ac$_3$	H	Acetyl	H	Acetyl	Acetyl
N-Acetyl-7,8,9-tri-*O*-acetylneuraminic acid	Neu5,7,8,9Ac$_4$	H	Acetyl	Acetyl	Acetyl	Acetyl
N-Acetyl-9-*O*-L-lactylneuraminic acid	Neu5Ac9Lt	H	Acetyl	H	H	L-Lactyl
N-Acetyl-4-*O*-acetyl-9-*O*-lactylneuraminic acid	Neu4,5Ac$_2$9Lt	Acetyl	Acetyl	H	H	Lactyl
N-Acetyl-8-*O*-methylneuraminic acid	Neu5Ac8Me	H	Acetyl	H	Methyl	H
N-Acetyl-8-*O*-sulfoneuraminic acid	Neu5Ac8S	H	Acetyl	H	Sulfate	H
N-Acetyl-9-*O*-phosphoneuraminic acid	Neu5Ac9P	H	Acetyl	H	H	Phosphate
N-Acetyl-2-deoxy-2,3-dehydroneuraminic acid	Neu5Ac2en	H	Acetyl	H	H	H
N-Glycolylneuraminic acid	Neu5Gc	H	Glycolyl	H	H	H
N-Glycolyl-4-*O*-acetylneuraminic acid	Neu4Ac5Gc	Acetyl	Glycolyl	H	H	H
N-Glycolyl-7-*O*-acetylneuraminic acid	Neu7Ac5Gc	H	Glycolyl	Acetyl	H	H
N-Glycolyl-9-*O*-acetylneuraminic acid	Neu9Ac5Gc	H	Glycolyl	H	H	Acetyl
N-Glycolyl-7,9-di-*O*-acetylneuraminic acid	Neu7,9Ac$_2$,5Gc	H	Glycolyl	Acetyl	H	Acetyl
N-Glycolyl-8,9-di-*O*-acetylneuraminic acid	Neu8,9Ac$_2$,5Gc	H	Glycolyl	H	Acetyl	Acetyl
N-Glycolyl-7,8,9-tri-*O*-acetylneuraminic acid	Neu7,8,9Ac$_3$,5Gc	H	Glycolyl	Acetyl	Acetyl	Acetyl
N-Glycolyl-8-*O*-methylneuraminic acid	Neu5Gc8Me	H	Glycolyl	H	Methyl	H
N-Glycolyl-8-*O*-sulphoneuraminic acid	Neu5Gc8S	H	Glycolyl	H	Sulfate	H

Source: Taken from Schauer, 1982.

At the present time, there are more than 20 known sialic acids (Table 3). The amino group bears either an acetyl or a glycolyl group. Almost the only substituent found in humans is the acetyl group. There is considerable variation in the hydroxyl substituents at C-4, C-5, C-7, C-8, and C-9. Acetyl, lactyl, methyl, sulfate and phosphate groups have been found (Table 3).

Distribution of sialic acids in nature

Sialic acids are found in all vertebrates (Mammalia, Aves, Reptilia, Amphibia, Pisces) and within each organism they are present in essentially all tissues (Warren, 1963; Corfield and Schauer, 1982) (Fig. 2). Sialic acids, mostly *N*-glycolylneuraminic acids, are also present in the cephalochordate *Amphioxus*, in the hemichordate *Dolichoglossus*, and in all five classes of echinoderms. They are not present in adult urochordates (tunicates) nor in their larvae, which contain a notochord that disappears in the adult. The absence of sialic acids in tunicates is surprising because they are believed to have arisen later in evolution than the echinoderms. Sialic acids are not present, or their distribution is sporadic, outside of these groups. For instance, they are not found in yeast, fungi, algae, lichens, or plants, but they are present in a few bacteria – some strains of *Escherichia coli* (K235), *Neisseria meningitidis* strain 1908, citrobacter, salmonella, and others. Bound sialic acids have not been found in amoeba, but appear to be present in a malarial parasite *Plasmodium berghei* (Sullivan and Volcani, 1974) a marine diatom (Seed *et al.*, 1974) and in *Trypanosoma cruzi* (Corfield and Schauer, 1982).

The erratic distribution of sialic acids among bacteria, and possibly certain protozoa, suggest that the enzymes responsible for their synthesis and metabolism were fortuitously acquired during association with animal cells. Sialic acid-containing bacteria are usually pathogenic (Irani and Ganapathi, 1962; Luppi and Cavazzini, 1966). The glycoproteins of some viruses such as vesicular stomatitis virus (VSV) (Hunt and Summers, 1976a,b), Rous sarcoma virus (Krantz, Lee and Hung, 1974), and rabies

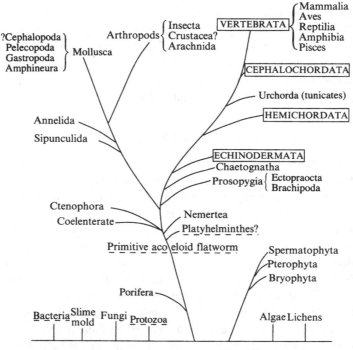

Fig. 2. Phylogenetic scheme showing location of sialic acids in nature. All of the groups listed above have been examined. Sialic acids are present in all of the species tested in the groups in boxes. The broken line signifies that only some species of the group contain sialic acids. The question mark signifies that only one species of the group tested was found to contain sialic acid, and it is not known whether the sialic acids are of endogenous origin.

(Dietzschold, 1977) contain sialic acids. However, the composition and structure of the carbohydrate components of viral glycoproteins are dependent on the type of cell in which the virus is grown. For instance, the glycoprotein of VSV virus grown in an animal cell contains sialic acids, while they are absent in that of virus cultivated in mosquito cells (Schloemer and Wagner, 1974,1975) which are devoid of sialic acids.

Sponges (Porifera), Prosopygia, Ctenophora and coelenterates, Annelida, Sipunculida Bryozoa and Nemertea are devoid of sialic acids. It would be of real interest to see whether these organisms have need of a molecule equivalent to sialic acids. Numerous species of arthropods (insects, spiders) and molluscs have been tested with negative results, except for the digestive gland of the American lobster (arthropod) and the squid (mollusc) (Warren, 1963). The sialic acids in these samples could be of exogenous origin. A particularly interesting occurrence is in a primitive platyhelminth, an acoel turbellarian, *Polychoerus carmelensis*, which is the most primitive metazoan known, perhaps arising from multinucleated ciliates. It has been suggested that the sialic acids are present in bacteria associated with the acoel turbellarian (Corfield and Schauer, 1982).

If we discount the presence of sialic acids in a few bacterial strains, it is likely that sialic acids appeared and evolved in Precambrian times, more than 5.5×10^8 years ago, well after the separation of plants from animals, since they are not found in plants. Sinohara (1972) has suggested that glycoproteins in which carbohydrate groups are linked via an N-asparagine linkage and which are particularly rich in terminal sialic acids arose later in evolution with the formation of tissues. Sialic acids, are present in all five classes of the invertebrate echinoderms and so may have arisen with this phylum: they are relative newcomers compared to other sugars. In this phylum, the N-glycolyl type is predominant. O-Acetylated forms of sialic acid (7-O and 9-O), rare forms such as 8-O-methyl N-glycolylneuraminic acid, and sialic acids bound in the interior part of sugar sequences (in glycolipids) rather than in the usual terminal position, are special characteristics that appear to be associated with the first occurrence of this family of carbohydrates. These unusual features would render them less susceptible to removal by sialidases.

The state of sialic acids in nature

Sialic acids almost always occur in bound form, although relatively large quantities of the free form have been found in the eggs of the eastern brook and rainbow trout (Warren, 1960). Approximately 50% of the total sialic acid content of the egg is in free form and this consists of both *N*-acetylneuraminic acid and *N*-glycolylneuraminic acid. The fraction of sialic acids that is free depends on the state of development of the embryo (Rahmann and Breer, 1976). Sialic acids in the free, reducing form have also been found in human cerebrospinal fluid (Saifer and Siegel, 1959; Papadopoulos and Hess, 1960; Jakoby and Warren, 1961) and hog stomach (Atterfelt *et al.*, 1958). Free sialic acids are found in urine, and in patients with the disease sialuria as much as 7 g of sialic acid can be eliminated in one day (Montreuil *et al.*, 1968). This is consistent with the finding that radioactive *N*-acetylneuraminic acid injected into rabbits appears almost immediately and completely in the urine. Apparently, the renal threshold is extremely low (Warren, unpublished).

Sialic acids are usually at the terminal position of sugar chains, where they are highly exposed and functionally important. Linkage is glycosidic (ketosidic), between the reducing position of the sialic acid at C-2 and an hydroxyl group of D-galactose, *N*-acetylglucosamine, *N*-acetylgalactosamine or another sialic acid residue. In the last case, the C-2 of one sialic acid molecule is bound to a hydroxyl group at C-8 on a second sialic acid residue. Polysialic acids with this linkage are found in glycolipids, adhesive glycoproteins of neural tissue and in polymers produced by certain strains of *E. coli* and *N. meningitidis*. Sialic acids are not only found in the bound state in macromolecules but they may be found also in relatively small, water soluble molecules such as neuraminlactose, and in unusual nucleotides such as UDP-*N*-acetylglucosamine-galactose-sialic acid. The function of the latter, found in goat colostrum, is unknown (Jourdian and Roseman, 1963).

The metabolism of hexosamines and sialic acids

Essentially, all of the reactions involved in the synthesis, degradation, modification, activation and transfer of hexosamines and sialic acids to macromolecules have been worked out, and these

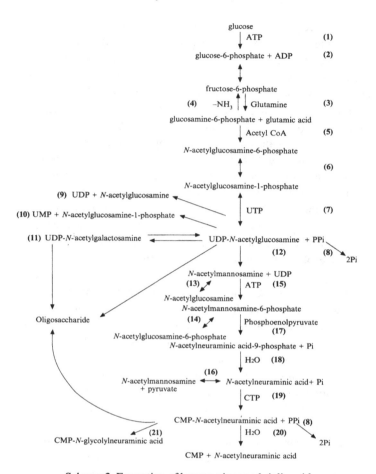

Scheme 2. Formation of hexosamines and sialic acids.

will be described below. Of special interest are the several devices, used by the cell, by which the levels of hexosamines and sialic acids, both free and bound, are controlled. They illustrate the incredible complexities of the cell and the fine balance necessary to maintain a steady state equilibrium. A defect in the mechanism can result in a serious disease such as sialuria.

The *N*-acetylhexosamine incorporated into sialic acid is derived from D-glucose via the glycolytic pathway intermediate fructose-6-phosphate. An outline of the many reactions involving the hexosamines and sialic acids is seen in Scheme 2. All nine carbon atoms of sialic acid are derived from glucose and find their way into neuraminic acid through the intermediates of glycolysis, fructose-1-phosphate (C4–9) and phosphoenolpyruvate (C1–3).

Glucose enters the glycolytic path by a hexokinase catalyzed phosphorylation at C-6 (Scheme 2, Reaction 1). The hexose becomes an aminosugar by the amination of fructose-6-phosphate catalyzed by glucosamine-6-phosphate synthetase (Reaction 2), mainly in the liver. The amino N is provided by the amide nitrogen of glutamine. A deaminase is capable of converting glucosamine-6-phosphate to fructose-6-phosphate and NH_3 (Reaction 4), but is incapable of acting on *N*-acetylglucosamine-6-phosphate. Thus, the acetylation of the amino group by acetyl CoA and a specific acetyl transferase for glucosamine-6-phosphate prevents the degradation of glucosamine-6-phosphate (Reaction 5) and permits the biosynthetic sequence to proceed. Acetylation removes the substrate of the deaminase (Comb and Roseman, 1958a). The acetyltransferase cannot transfer the glycolyl group of glycolyl CoA to glucosamine-6-phosphate. Although acetylation spares glucosamine-6-phosphate from destruction, the deaminase is stimulated by *N*-acetylglucosamine-6-phosphate, the product of acetylation, thus preventing an overproduction of hexosamine. While synthesis of hexosamine via the glutamine pathway is the primary reaction, glucosamine-6-phosphate can also be synthesized by amination of fructose-6-phosphate, i.e. by reversal of the deamination reaction. However, equilibrium strongly favors deamination and the enzyme is considered degradative.

In order for free *N*-acetylglucosamine-6-phosphate to be metabolized further so that it can be polymerized and incorporated into glycoproteins, or be converted to other essential aminosugars, it must be activated to form a nucleoside diphosphate-*N*-acetylglu-

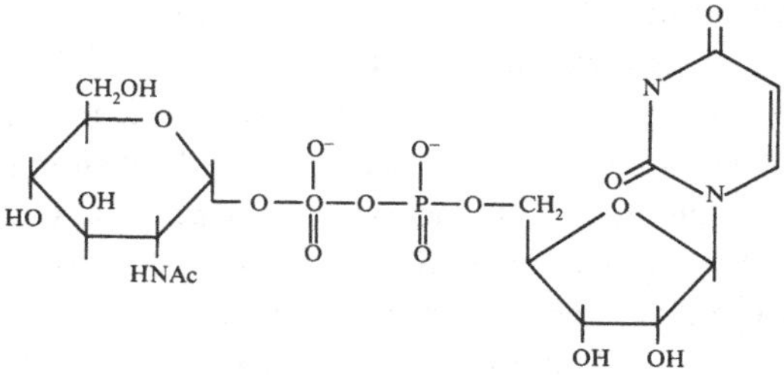

Scheme 3. UDP-*N*-acetylglucosamine.

cosamine (Scheme 3). For the activating pyrophosphorylase reaction to occur (Reaction 7), there must be a migration of phosphate from the 6-hydroxyl to the 1-hydroxyl position (Reaction 6).

$$N\text{-acetylglucosamine-6-phosphate} \xleftrightarrow{\text{Mg}^{2+}}$$
$$N\text{-acetylglucosamine-1-phosphate} \quad (6)$$

The reaction is catalyzed by a widely distributed mutase which is distinct from phosphoglucomutase of the glycolytic cycle. The latter enzyme can catalyze Reaction 6 but only at a very slow rate. Enzyme activity is dependent upon the presence of catalytic quantities of either glucose-1,6-diphosphate or *N*-acetylglucosamine-1,6-diphosphate. Although at equilibrium the 6-phosphate form is favored (84% vs 14% for the 1-phosphate), the pyrophosphorylase-catalyzed reaction utilizes the 1-phosphate and 'pulls' the reaction forward, impelled by the irreversible hydrolysis of the product inorganic pyrophosphate (PP_i), by inorganic pyrophosphatase (Reaction 8).

$$N\text{-acetylglucosamine-1-phosphate} + UTP \xleftrightarrow{\text{Mg}^{2+}}$$
$$UDP\text{-}N\text{-acetylglucosamine} + PP_i \quad (7)$$

$$PP_i \xrightarrow{H_2O} 2 P_i \quad\quad\quad\quad\quad (8)$$

The key intermediate, UDP-*N*-acetylglucosamine, can take part in several important reactions. It can be hydrolyzed to form UDP +

N-acetylglucosamine (Reaction 9). A pyrophosphatase can cleave the pyrophosphate linkage of UDP-*N*-acetylglucosamine to yield UMP and *N*-acetylglucosamine-1-phosphate (Reaction 10). UDP-*N*-acetylglucosamine is the essential 'activated' sugar in polymerization reactions, and it can undergo epimerization at C-4 to form UDP-*N*-acetylgalactosamine (Reaction 11). Finally, UDP-*N*-acetylglucosamine can undergo a complex and irreversible epimerization at C-2, accompanied by hydrolysis at C-1, to form *N*-acetylmannosamine and UDP (Reaction 12). The labile enzyme that catalyzes the reaction requires no coenzymes or factors.

$$\text{UDP-}N\text{-acetylglucosamine} \xrightarrow{\text{H}_2\text{O}} N\text{-acetylmannosamine} + \text{UDP} \quad (12)$$

N-Acetylmannosamine can also be formed directly from *N*-acetylglucosamine (Ghosh and Roseman, 1965a) (Reaction 13).

$$N\text{-acetylglucosamine} \longleftrightarrow N\text{-acetylmannosamine} \quad (13)$$

It is possible that the function of the enzyme catalyzing this reaction is to rid the cell of excess *N*-acetylmannosamine and thus regulate the level of sialic acid biosynthesis. In bacteria, a similar reaction can occur except that the *N*-acetylhexosamine-6-phosphate derivatives are interconverted (Ghosh and Roseman, 1965a,b) (Reaction 14).

N-Acetylmannosamine is of special importance because it is the first product in the sequence that is specific for the sialic acid pathway. In its conversion to sialic acids in animal cells, it is phosphorylated by a kinase (Warren and Felsenfeld, 1962) to form *N*-acetylmannosamine-6-phosphate (Reaction 15).

$$N\text{-acylmannosamine} + \text{ATP} \xrightarrow[\text{K}^+]{\text{Mg}^{2+}} N\text{-acylmannosamine-6-phosphate}$$
$$\text{(Acyl: acetyl or glycolyl)} \quad (15)$$

In 1956, Heimer and Meyer discovered an enzyme in extracts of *Vibrio cholerae* that could cleave sialic acid to an *N*-acylhex-

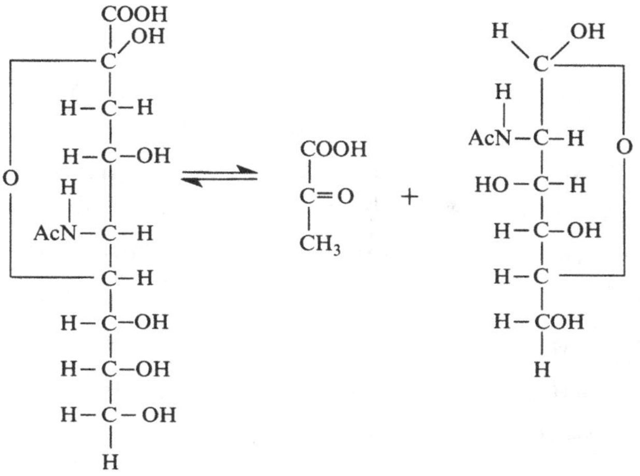

N-acetylneuraminic acid   Pyruvate   N-acetylmannosamine

Scheme 4. Cleavage of sialic acid.

osamine and pyruvate. The N-acylhexosamine was shown to be N-acetylmannosamine, (Comb and Roseman, 1958b, 1960) (Scheme 2, Reaction 16 and Scheme 4).

N-acetylneuraminic acid $\longleftrightarrow$ pyruvate + N-acetylmannosamine   (16)

Although the enzyme, sialic acid aldolase (N-acylneuraminate lyase), could operate in the direction of synthesis, equilibrium strongly favors degradation ($K = 0.064$). Further, the aldolase is present in sites where sialic acid can not be found, and is absent in tissues rich in sialic acids. It was believed to be a degradative, scavenging enzyme, and so other pathways of synthesis were sought. It was found that N-acetylmannosamine-6-phosphate could condense with the intermediate of glycolysis phospho-enolpyruvate (PEP) to form N-acetylneuraminic acid-9-phosphate (Roseman *et al.*, 1961; Warren and Felsenfeld, 1962) (Scheme 2, Reaction 17 and Scheme 5). The enzyme catalyzing this irreversible reaction was called sialic acid-9-phosphate synthetase.

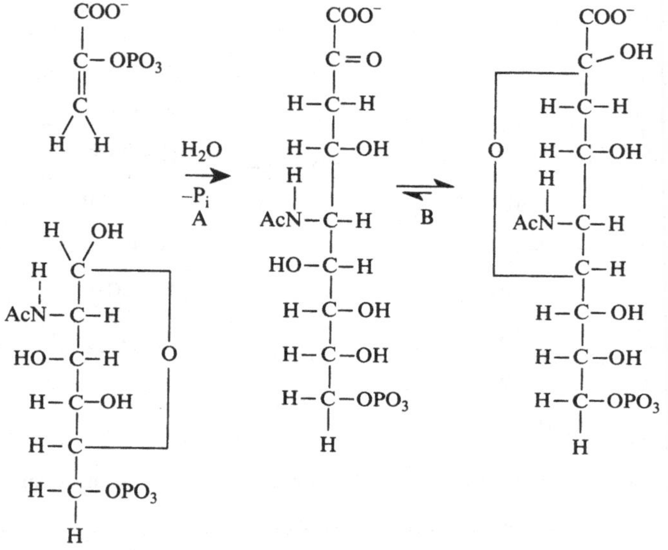

Scheme 5. Phosphoenolpyruvate reacts with N-acetylmannosamine to form N-acetylneuraminic acid-9-phosphate.

$$N\text{-acetylmannosamine-6-phosphate} + PEP \xrightarrow{H_2O} $$
$$N\text{-acetylneuraminic acid-9-phosphate} + P_i \quad (17)$$

Although non-specific phosphatases can remove the 9-phosphate, a specific sialic acid-9-phosphate phosphatase does exist that catalyzes the dephosphorylation (Reaction 18), to form the substrate for the activation reaction described below (Reaction 19).

$$N\text{-acetylneuraminic acid-9-phosphate} + H_2O \rightarrow$$
$$N\text{-acetylneuraminic acid} + P_i \quad (18)$$

Of the various reactions on the pathway to the formation of sialic acid there are three that are essentially irreversible: the formation of N-acetylmannosamine (Reaction 12), the condensation of phosphoenol pyruvate with N-acetylmannosamine-6-phosphate (Reaction 17), and the dephosphorylation of sialic acid-9-

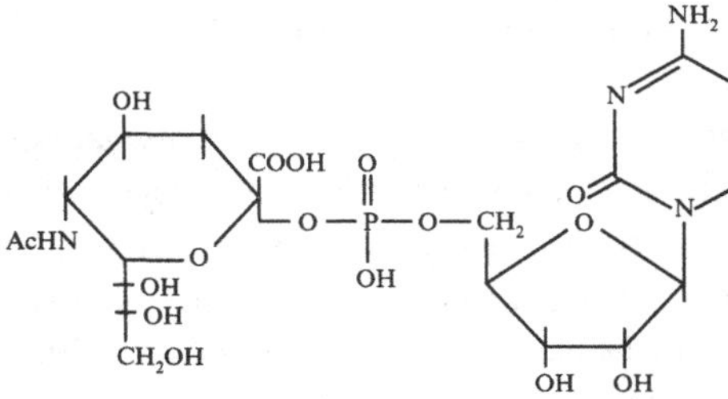

Scheme 6. CMP-*N*-acetylneuraminic acid (CMPNeu5Ac).

phosphate (Reaction 18). Not only is the formation of sialic acid favored, but activation to form CMP-sialic acid is, for all practical purposes, also irreversible. However, there are various isomerases, phosphatases and pyrophosphatases in the cell that can destroy intermediates and thereby reduce synthesis of activated sialic acid.

It is odd that in a bacteria such as *Neisseria meningitidis*, *N*-acetylmannosamine is formed by the epimerization of *N*-acetylglucosamine-6-phosphate, and the product, *N*-acetylmannosamine-9-phosphate, must be dephosphorylated for its incorporation into sialic acid:

$$N\text{-acetylmannosamine} + PEP \xrightarrow{\text{H}_2\text{O}} N\text{-acetylneuraminic acid} + P_i$$

By comparison in the animal cell, the dephosphorylated hexosamine *N*-acetylmannosamine is formed (Reaction 12), and it must be phosphorylated (Reaction 15) for conversion to sialic acid (Blacklow and Warren, 1962) (Reaction 17). In each system an 'extra' step in the sequence is inserted, probably as part of a control mechanism.

Cytidine-5'-monophospho-*N*-acetylneuraminic acid (CMPNeu 5Ac) (Scheme 6), the activated form of sialic acid, is synthesized by a pyrophosphorylase type of reaction catalyzed by CMP-sialic acid synthetase (Roseman, 1962; Warren and Blacklow, 1962) (Reaction 19).

$$\text{Neu5Ac+CTP} \xrightarrow{\text{Mg}^{2+}} \text{CMPNeu5Ac+PP}_i \qquad (19)$$

In animal tissues, both N-acetyl- and N-glycolylneuraminic acid as well as their acetylated derivatives N-acetyl-7-O-acetylneuraminic acid, N-acetyl-9-O-acetylneuraminic acid, and N-acetyl-4-O-acetylneuraminic acid, can be activated,. The reaction is slightly reversible, while in the bacterium *N. meningitidis* it is irreversible.

Although the reaction appears to resemble that catalyzed by a pyrophosphorylase, it is quite different. It is, in fact, unique, differing from the biosynthesis of a typical activated sugar, e.g. UDP-N-acetylglucosamine (Reaction 7), which is fully reversible. Here the UTP reacts with the sugar-1-phosphate to form a product with a pyrophosphate link, while in the synthesis of CMPNeu5Ac there is direct interaction between the inner phosphate of the nucleotide triphosphate CTP and the free, reducing group of Neu5Ac. Unlike other activated sugars (UDP-glucose, GDP-mannose, UDP-N-acetylhexosamine), CMPNeu5Ac contains a single phosphate rather than a pyrophosphate bridge.

A hydrolase has been found that irreversibly converts CMPNeu5Ac to CMP + Neu5Ac (Reaction 20), while a pyrophosphatase splits all the other activated sugars into a sugar-1-phosphate, and a nucleotide monophosphate (Reaction 10). These enzymes appear to be concentrated in the plasma membrane.

$$\text{UDP-sugar} \xrightarrow{\text{H}_2\text{O}} \text{UMP} + \text{sugar-1-phosphate}$$

The conversion of N-acetyl (Ac) to N-glycolyl (Gc) groups, catalyzed by a hydroxylase occurs in a variety of tissues at the level of activated sialic acid (Shaw and Schauer, 1988) (Reaction 21).

$$\text{CMPNeu5Ac} \rightarrow \text{CMPNeu5Gc} \qquad (21)$$

The reaction catalyzed by a soluble monooxygenase requires NADPH, molecular oxygen and ferrous ion. Sialic acid synthetase of animal tissue can use N-glycolylmannosamine to form N-glycolylneuraminic acid and this sialic acid can also be activated by a CMP-sialic acid synthetase.

Metabolic control of hexosamines and sialic acids

We have discussed the pathways by which hexosamines and sialic acids are derived from glycolytic intermediates in preparation for their incorporation into glycolipids and the glycoconjugates. It has taken almost half a century to delineate these pathways. However, the living cell can not only carry out these reactions, it also can, in sensitive ways, regulate the rates at which intermediates are produced, destroyed or incorporated into larger molecules. Over vast periods of time, subtle and intricate mechanisms have come into play, so that an integrated steady state has evolved in which various intermediates are present, when required, at the right place and at the right concentrations. The metabolism of the hexosamines and sialic acids, just discussed, is characterized by a number of known regulating devices, and these changes can vary with time during development of the cell and organism. There appears to be a balance of synthesis and degradation of metabolites and ultimate products. Regulation is effected in many ways:

(a) While hexosamine phosphate intermediates are made through kinase activities, there are numerous degradative phosphatases, both non-specific and specific, that reduce the concentrations of these intermediates; a balance must be established.

(b) The amination of fructose-6-phosphate (Reaction 3) is the first step specific for hexosamine synthesis leading to the formation of UDP-*N*-acetylglucosamine. The amination reaction is inhibited effectively by this activated sugar (Kornfeld *et al.*, 1964), which in a sense is the final product of the sequence. This is a classic example of feedback inhibition.

(c) The level of glucosamine-6-phosphate can also be lowered by deamination (Reaction 4). However, as discussed, deamination can be prevented by transfer of an acetyl group to the amino group, and so a competition exists between the acetylase and the deaminase to reduce the level of glucosamine-6-phosphate. To complicate matters, the product of the acetylase, *N*-acetylglucosamine-6-phosphate, activates the deaminase, although it does not take part in the reaction (Pattabiraman and Bachhawat, 1961). It is apparent that the concentrations of intermediates are tightly regulated.

(d) The conversion of UDP-*N*-acetylglucosamine to mannosamine is the first step committed to the formation of sialic acid, while CMP-sialic acid is the final product of the sequence, prior to transfer of sialic acid to various lipids or glycoproteins. As another example of feedback inhibition, it has been found that CMP-sialic acid is an effective inhibitor of UDP-*N*-acetylglucosamine 2'-epimerase, the enzyme that forms mannosamine, the first product specific for sialic acid metabolism (Reaction 12).

(e) The level of *N*-acetylmannosamine can be reduced by a 2'-epimerase that can convert it to *N*-acetylglucosamine (Reaction 13), lowering its level and thereby reducing the level of sialic acid synthesis.

(f) Subcellular localization of enzymes must also be important in the ebb and flow of metabolites. The enzyme responsible for the formation of CMP-sialic acid, CMP-sialic acid synthetase, is located in the nucleus (Kean, 1991) where it is unlikely that glycoprotein synthesis occurs. Synthesis of sialic acid, as well as the other activated sugars, by soluble enzymes takes place in the cytoplasm. Sialic acid must enter the nucleus, where it is activated. CMP-sialic acid leaves the nucleus and, at least in part, enters the Golgi apparatus, the site of terminal glycosylation of glycoproteins by sialyl transferases. Passage of the CMP-sialic acid across the Golgi membrane into the lumen is effected by a translocase that can be inhibited by 5'-CMP and 5'-UMP because, as one mole of CMP-sialic acid enters the lumen, a mole of CMP leaves (Zapata *et al.*, 1989; Schauer, 1991). 5'-CMP is one of the products after transfer of sialic acid as well as a product of CMP-sialic acid hydrolysis.

CMP-sialic acid can also be degraded by a hydrolase that is found largely, if not totally, in the surface membrane of the cell. The location of this degradative enzyme seems remarkable in view of the fact that CMP-sialic acid is made in the nucleus. The hydrolase is inhibited by CTP and other nucleotides. Sulfhydryl-containing compounds also inhibit the hydrolase, while they stabilize the CMP-sialic acid synthetase. UDP-*N*-acetylglucosamine is also an inhibitor of CMP-sialic acid hydrolase. In addition, there are nucleotide pyrophosphatases that act upon other activated sugars with their own sets of regulating devices. In summary, the level of CMP-sialic acid is obviously an important determinant of the biosynthetic capacity of the cell, depending on a delicate and complex balance of synthesis, degradation and utilization, each of which takes place at a different site in the cell. By a feedback inhibition process, it has some control of its own synthesis and, in addition, several other related metabolites exert their control of its synthesis and degradation.

(g) The activated sugars are made in the cytoplasm and must be transported through the Golgi and endoplasmic reticulum (ER) membranes to the site of sugar transferase activity. There is growing evidence that entry of an activated sugar molecule is balanced by exit of a nucleoside monophosphate by an antiport transport mechanism that is electroneutral: one molecule of GMP leaves the Golgi lumen for every molecule of GDP-fucose that enters, without expenditure of energy. Defects in this mechanism drastically reduce glycoprotein biosynthesis because of a shortage of activated sugars at the site of oligosaccharide assembly (Hirschberg, 1987).

(h) The regulation of gene expression of every enzyme in hexosamine and sialic acid metabolism, including the sugar transferases, is critically important. The levels of sialyl transferases may be of special significance in glycoprotein biosynthesis because, when a sialic acid is added, no further chain elongation occurs: the sialic acid is a 'capper' or chain terminator. Thus, the level of sialyl transferase(s) may determine the size of the bound sugar groups.

(i)   In general, every intermediate structure formed may be competed for by more than one enzyme. Therefore, its subsequent history may depend on its concentration and location within the cell, the amounts of each of the competing enzymes and their kinetic characteristics (pH optimum, the shape of the pH curve, K_m value, inorganic ion requirements), their response to hormones, and conditions within the cell. The levels of nucleotides in the cell could be important. ATP and other nucleoside triphosphates are required in stoichiometric amounts, while others affect the rates of reactions; for instance CTP has been shown to inhibit sialyl transferases. In numerous cases, successful competition leads to progression along a biosynthetic pathway: *N*-acetylation *vs* deamination of glucosamine-6-phosphate; the rate of biosynthesis of sialic acid *vs* the rate of its activation *vs* the rate of its degradation by sialic acid aldolase. In other situations, there is reduction in the amount of intermediates by dephosphorylation or hydrolysis. A steady-state evolves and is maintained in these systems.

The carbohydrate groups attached to proteins and lipids are synthesized, one monomer at a time, by transfer of sugars from their activated form (UDP-galactose, UDP-*N*-acetylglucosamine, UDP-*N*-acetylgalactosamine, GDP-mannose, GDP-fucose, CMP-sialic acid) to the already formed carbohydrate polymer. The product of one sugar transferase reaction becomes the substrate (sugar recipient) of the next. Transferases embedded in the Golgi membrane must be relatively fixed and bear a defined spacial relationship with one another. A basis for subtle biosynthetic control may be provided depending on the state of the membrane, the availability of each type of activated sugar, and the environmental conditions (pH, temperature, inorganic ion concentrations, oxidation–reduction state of the cell).

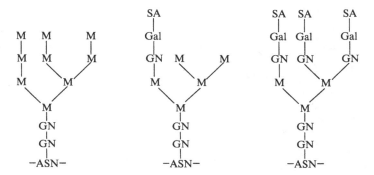

Scheme 7. Three major classes of oligosaccharide structures bound to protein through the amide N of asparagine. ASN, asparagine; GN, *N*-acetylglucosamine; M, mannose; Gal, galactose; SA, sialic acid.

The level of expression of each transferase in this membrane-bound constellation is important and, in recent years, the subject has been studied intensively. In addition, the cloning of genes and analysis of their expression has led to the complete sequencing of enzymes such as CMP-sialic acid synthetase (Zapata *et al.*, 1989), a virtually impossible task in the mid-1980s. The gene for this enzyme, when inserted into the genome of *E. coli,* is over-expressed so that 10% of the soluble protein is CMP-sialic acid synthetase (Shames *et al.*, 1991).

Asparagine *N*-glycosylation

The most abundant form of protein-bound oligosaccharides are those linked to the amide nitrogen of asparagine, and among these there are three major classes of oligosaccharide structures whose biosynthesis will be discussed below (Scheme 7).

There are a large number of variations on these basic structures. Two are of special note: one bearing a polysialic acid and another bearing a polylactosamine molecule. Dolichol, a widely distributed lipid, is formed by the head to tail condensation of 19

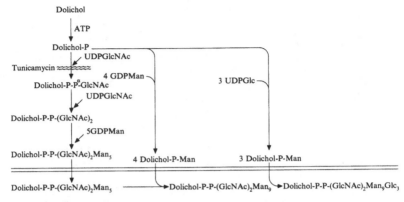

Scheme 8. Formation of an oligosaccharide-lipid intermediate for the *N*-glycosylation of asparagine. The reactions above the double line (membrane) take place on the cytoplasmic side of the rough endoplasmic reticulum (ER); those below on the lumenal side. Note that at a certain stage the dolichol-P-P-saccharides traverse the membrane, and the last seven sugars are added within the lumen of the ER. GlcNAc, *N*-acetyl-D-glucosamine; Man, D-mannose; Glc, D-glucose; P, phosphate.

isopentenyl pyrophosphate units. The terminal hydroxyl is phosphorylated by a dolichol kinase and ATP is the phosphate donor.

The molecule is about 10 nm long, which is considerably greater than the thickness of a lipid bilayer with which it is associated during its functioning. It is likely that it wraps itself around or covers the saccharide component to facilitate traversal of cytoplasmic membranes.

The *N*-glycosylation pathway begins with the stepwise formation of an oligosaccharide-lipid intermediate, dolichol-P-P-(GlcNAc)$_2$-Man$_9$Glc$_3$, according to Scheme 8 (Struck and Lennarz, 1980).

The initial step specific for glycosylation of asparagine is the synthesis of dolichol-P-P-GlcNAc, catalyzed by the enzyme UDPGlcNAc:dolichol-phosphate-*N*-acetylglucosamine-1-phosphate transferase.

UDP-GlcNAc + dolichol-phosphate → dolichol-P-P-GlcNAc + UMP

The enzyme is located in the ER. It should be noted that glucosamine-1-phosphate is transferred rather than glucosamine itself, as occurs in the usual sugar transfers catalyzed by glycosyl transferases. GlcNAc is bound to phosphate in UDPGlcNAc by an α-linkage and this configuration is maintained in dolichol-P-P-GlcNAc. In the course of other transfers of GlcNAc from UDPGlcNAc to acceptor, an inversion takes place, so that the linkage between C-1 of GlcNAc and the acceptor hydroxyl is β. In the synthesis of dolichol-P-P-GlcNAc, a pyrophosphate linkage is broken and another is made. Since it is the first step in the important process of asparagine glycosylation, the responsible enzyme should be especially susceptible to numerous intracellular controls (Lehrman, 1991). It is strongly inhibited by the relatively specific drug tunicamycin, derived from *Streptomyces lysosuperficus*. Since inhibition occurs at an early stage in the stepwise synthesis, the drug is capable of preventing the formation of most dolichol oligosaccharide intermediates and effectively arrests glycosylation of asparagine residues. GlcNAc, mannose and glucose molecules are added sequentially to dolichol-P-P-GlcNAc, both on the cytoplasmic and the luminal side of the ER membrane. Of the nine mannose residues in the complete oligosaccharide, the first five are derived from GDP-mannose directly, and this occurs on the cytoplasmic face of the ER. The terminal four mannose and three glucose residues are derived from dolichol-P-mannose and dolichol-P-glucose, and are transferred to the dolichol-oligosaccharide within the lumen of the ER. It would appear that the sugars do not traverse the ER membrane in their activated nucleotide form, but, when attached to dolichol and surrounded by this lipophilic molecule, passage through the membrane can take place.

The oligosaccharide $(GlcNAc)_2$-Man_9-Glc_3 is transferred en bloc from the dolichol-P-P carrier to the amide N of asparagine of the nascent polypeptide within the lumen of the rough ER. The recipient asparagine residue must be a component of a tripeptide, ASN-X-Ser/Thr, although this sequence alone is not sufficient for reception to take place (X can be virtually any amino acid). There

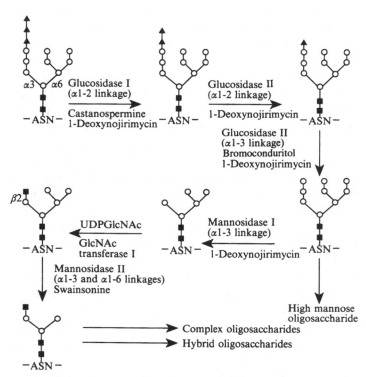

Scheme 9. Early processing of protein-bound carbohydrates. ASN asparagine; ■ GlcNAc; ○, mannose; ▲, glucose.

must be other steric factors which determine whether the oligosaccharide will be transferred to a specific asparagine. Transfer takes place as the polypeptide is being synthesized and is emerging on the luminal surface of the rough ER. Since transfer of sugars takes place before completion of the polypeptide, glycosylation may be considered a co-translational as well as a post-translational event. Processing of the oligosaccharide on the asparagine amide *N* then occurs (Struck and Lennarz, 1980; Kobata, 1984; Kornfeld and Kornfeld, 1985; Schacter, 1991) (Scheme 9).

A terminal glucose residue is removed by α-glucosidase I, an enzyme that is inhibited by 1-deoxynojirimycin and castanospermine. The two remaining glucose residues are cleaved separately by α-glucosidase II which can be inhibited by 1-deoxynojirimycin and bromoconduritol. A single mannose residue is removed by α-mannosidase I, which is an enzyme inhibited by 1-deoxymannojirimycin. As the processed molecule is transferred to the cis-Golgi by a vesicular transport system, three more mannose residues are removed by α-mannosidase I. Four mannose and all three glucose residues have been removed. In the medial-Golgi, GlcNAc is transferred from UDPGlcNAc to the α1-3 mannose in a β1-2 linkage by *N*-acetylglucosaminyl transferase I. Once this critical step occurs, the product can be acted upon by five different enzymes:

(a) α-Mannosidase II removes two mannose residues and leads to the synthesis of complex oligosaccharides. The GlcNAc residue must be present before α-mannosidase II can act. This enzyme is inhibited by swainsonine.

(b) GlcNAc transferase II transfers GlcNAc from UDPGlcNAc to the (α1-6) mannose to yield a product that can be acted upon by several other enzymes to form different complex structures such as bi-antennary, bisected bi-antennary, tri-antennary, and complex tri-antennary structures as seen in Scheme 10.

(c) GlcNAc transferase III activity is responsible for the formation of 'bisected' structures by transferring GlcNAc to the 4'-hydroxyl of the inner mannose in a β-linkage.

(d) GlcNAc transferase IV catalyzes the transfer of GlcNAc to the 4'-hydroxyl group of the core α1-3 or α1-6 mannose that leads to the formation of the tri-antennary or bisected tri-antennary structures by the addition of galactose and sialic acid residues.

(e) Fucosyl transferase is responsible for the transfer of fucose from GDP-fucose to the core GlcNAc, which is attached to the asparagine amide group. It should be noted that this GlcNAc residue is remote from the GlcNAc residue attached to mannose necessary for fucosyl transferase activity.

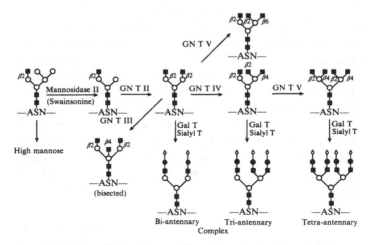

Scheme 10. Later processing of protein-bound carbohydrates. ASN asparagine; ■ GlcNAc; ○, mannose; ●, galactose, ◊, sialic acid; GN T, GlcNAc transferase; Gal T, galactose transferase; sialyl T, sialyl transferase.

Numerous sequences have been described and several branch points have been defined. The final pathway taken depends on the relative activities of the various enzymes involved (Scheme 10). A cell that has a low level of α-mannosidase II but a high GlcNAc transferase III activity will synthesize more bisected, hybrid oligosaccharides. GlcNAc transferase I is therefore a key enzyme that forms a product which is the substrate for at least five other enzymes. The absence of this enzyme would abolish the formation of complex and hybrid oligosaccharides, while permitting the synthesis of high-mannose structures.

A bisecting GlcNAc residue in an oligosaccharide prevents it from being a substrate for α-mannosidase II, fucosyl transferase or GlcNAc transferases IV and V. A critical competition exists between GlcNAc transferase III and GlcNAc transferases IV and V for bi-antennary acceptor. Addition of galactose in the trans-Golgi

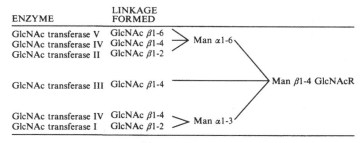

ENZYME	LINKAGE FORMED
GlcNAc transferase V	GlcNAc β1-6
GlcNAc transferase IV	GlcNAc β1-4
GlcNAc transferase II	GlcNAc β1-2
GlcNAc transferase III	GlcNAc β1-4
GlcNAc transferase IV	GlcNAc β1-4
GlcNAc transferase I	GlcNAc β1-2

Scheme 11. Specificity of the GlcNAc transferases. R, GlcNAc-ASN.

inhibits the activity of GlcNAc transferases III, IV and V. These are located in the median-Golgi. Therefore, the sequence of addition of sugars can be very important in the determination of the final composition and structure of carbohydrate groups (Schacter, 1991).

In summary, the oligosaccharides are synthesized by the addition of sugars, one at a time, in succession, by reactions catalyzed by sugar transferases. The process is not guided by a template but by the inherent specificity of the enzyme (Scheme 11). It is probable that each enzyme is responsible for only one linkage and specificity depends on the sugar, its activated form, and the acceptor. It is not known, at present, how much the amino acid sequence around, or even distant from, the glycosylated amino acid influences the final form of the carbohydrate polymer. As we saw, one glycosylation step can determine whether a subsequent addition of a sugar can take place directly to the first sugar, or to another, remote sugar in the oligosaccharide. The amount of activated sugar may be an important factor, as well as the level of a specific sugar transferase activity, which is genetically determined, and may vary during embryonic development or may change in malignancy. This is true of oligosaccharide assembly of glycoproteins and glycolipids. The determinations of pathways have been greatly assisted by exploiting mutant cells that lack one or another sugar transferase. Drugs such as tunicamycin, swainsonine,

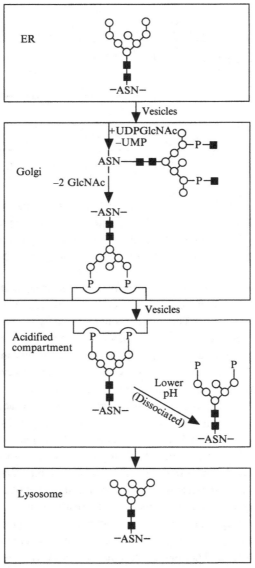

Scheme 12. Processing of lysosomal enzymes. ASN, asparagine; ■ GlcNAc; ○, mannose.

1-deoxynojirimycin and others that are relatively specific inhibitors of various processing enzymes, have also proven useful.

Processing of lysosomal enzymes

The scheme just described accounts for the post- and co-translational modifications of glycoproteins that are found in the lumen of the rough ER and Golgi apparatus. Most glycoproteins are eventually secreted or end up in the plasma membranes or in the lysosomes. The signal for targeting certain glycoproteins destined for the lysosomes is phosphorylated mannose in the protein-bound saccharide. The formation of protein-bound mannose-6-phosphate is catalyzed by a phosphotransferase in the Golgi apparatus (Kornfeld, 1987) which transfers N-acetylglucosamine-1-phosphate from UDP-GlcNAc to the 6-hydroxyl groups of certain, specific mannose residues in the glycoprotein to form a phosphodiester intermediate. The condensation produce is later hydrolyzed to form the bound mannose-6-phosphate and free GlcNAc by N-acetylglucosamine-1-phosphodiester α-N-acetylglucosaminidase.

$$\text{UDPGlcNAc} + \text{mannose-R} \rightarrow \text{GlcNAc-P-mannose-R} + \text{UMP}$$

$$\text{GlcNAc-P-mannose-R} \overset{\text{H}_2\text{O}}{\rightarrow} \text{GlcNAc} + \text{P-mannose-R}$$

The signal that specifies which mannose residues are phosphorylated by this unique reaction is unknown, but it probably resides in the amino acid sequence of all lysosomal glycoprotein enzymes and is highly specific. Once the mannose-6-phosphate recognition signal is formed, the glycoprotein enzymes are bound by one of two forms of specific receptor. Glycoproteins bearing oligosaccharides with two phosphomannose groups are favored over those with only one. The receptors carry the enzymes in small transport vesicles from the trans-Golgi to a prelysosomal compartment where they dissociate from the mannose-6-phosphate receptors on the membrane at the relatively low pH which exists within the vesicles. The receptors are recycled back to the Golgi while the enzymes are dephosphorylated and are transferred to the lysosomes. By this mechanism, occurring in the trans-Golgi, a special group of glycoproteins have been separated from the glycoproteins destined for secretion and those destined for the plasma membrane (Scheme 12).

There are numerous genetic diseases in humans that involve the synthesis, transport, packaging, and secretion of lysosomal enzymes. In I cell disease and pseudo-Hurler polydystrophy there is a deficiency of the phosphotransferase. Fibroblasts from patients cannot retain their lysosomal enzymes because bound mannose is not phosphorylated. They are not targeted for the lysosomes, and so are secreted into culture medium like other secretable glycoproteins (Kornfeld, 1986)

Biosynthesis of *O*-linked oligosaccharides

Carbohydrate groups can be linked to the hydroxyl groups of serine or threonine, and hydroxylysine. In mucins, GalNAc is linked to serine or threonine via an α-glycosidic bond, and the final oligosaccharide may only be a disaccharide, as in ovine submaxillary gland mucin. However, the carbohydrate groups of mucins in other sites, such as in ovarian cyst fluid, may be quite large and complex. In proteoglycans, sugar groups are linked to the polypeptide via a serine-xylose bonding, while in collagens the linking group is hydroxylysine-galactose.

The initiation of mucin biosynthesis (Schacter and Roseman, 1980) is carried out by a membrane-bound enzyme that transfers GalNAc from UDP-GalNAc to the hydroxyl groups of serine or threonine in a polypeptide (Scheme 13).

$$\text{UDPGalNAc} + \text{serine} \overset{R}{\underset{R}{<}} \longrightarrow \text{GalNAc-serine} \overset{R}{\underset{R}{<}} + \text{UDP}$$

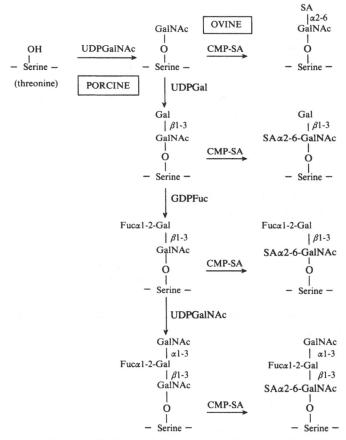

Scheme 13. Biosynthetic sequences for submaxillary gland mucin in sheep and pig. Fuc fucose; SA, sialic acid.

The transferase favors larger polypeptide acceptors. In contrast to transfer of carbohydrates to the asparagine amide N, where the triplet Asp-X-Ser (Thr) in the polypeptide is an absolute requirement, no specific sequences around an accepting hydroxyamino acid are known that render the amino acid a better acceptor.

In ovine submaxillary glands, the transfer of GalNAc to polypeptide is followed by sialylation of GalNAc at the 6-hydroxyl group to form an α2-6 linkage. Since sialic acid is almost always a terminal sugar, or renders a saccharide a poor acceptor, further additions of sugars do not take place, and so the ovine mucin molecule bears large numbers of sialyl α2-6 GalNAc disaccharide residues. This pathway is dominant in the ovine submaxillary gland because the gland is rich in CMP-sialic:GalNAc-mucin α 2-6 sialyl transferase. On the other hand, porcine submaxillary gland is rich in UDP-gal:GalNAc-β1-3 galactosyl transferase, which is responsible for the synthesis of a bound Gal-GalNAc disaccharide, that can be further glycosylated according to Scheme 13. A fucose or a GalNAc molecule can be further added by appropriate, specific transferases so that the sugar groups on porcine mucin are larger than those of ovine mucin. At any point during chain growth, once sialic acid is added, no further additions of sugar (fucose or GalNAc) can be made. In these examples, once the first sugar, GalNAc, is added, the future pathway is largely determined by competition between transferases for sialic acid and galactose. The relative amounts of the enzymes and of their substrates, UDP-Gal or CMP-sialic acid, and conditions within the cell determine which pathway is preferred.

Modification of sialic acids

The various enzymes that use sialic acids or their derivatives as substrates may or may not act on the same compounds bearing acetyl groups on C-4, -7, -8, -9. Substitutions on the hydroxyl groups alter the size, charge and hydrophobicity of the molecule. Sialyl acetylases, in the Golgi vesicles of rat liver, transfer the acetyl groups from acetyl CoA to a carrier in the membrane. One or more histidine residues are critically involved. In a second step, the acetyl group is transferred to hydroxyl groups of sialic acids of glycoproteins on the luminal side of the membrane (Higa *et al.*, 1989). Two specific acetylases are known: one transfers an acetyl group to the 4'-hydroxyl and another to the 8'-hydroxyl and 9'-hydroxyl positions of Neu5Ac.

Generally, human tissues contain only Neu5Ac but, in human melanoma, the ganglioside GD3 bears *N*-acetyl-9-*O*-acetylneuraminic acid. A monoclonal antibody has been produced against this structure that is partly effective in treating the tumor (Koprowski, 1987). Not only are there specific *O*-acetyl transferases, acting on both free and bound sialic acids, there are also esterases that remove acetyl groups. Lactyl, methyl, sulfate and phosphate groups may also be present (Table 3). The methyl group of 8-*O*-methyl-*N*-glycolylneuraminic acid, found in starfish (Warren, 1964) is synthesized by the transfer of the methyl group from *S*-adenosylmethionine to bound, but not free, *N*-acetyl and *N*-glycolylneuraminic acid (Herrier *et al.*, 1985).

While specific functions cannot be ascribed to these substituents, it is known that macromolecules bearing acetylated sialic acids are rendered resistant to degradation by neuraminidase, thus increasing the half-life of serum glycoproteins and glycoproteins on cell surfaces. The presence of *O*-acetyl groups appears to be programed during embryonic development and in tissue-specific expression. They can reappear and are detected as oncofetal antigen determinants (Thurin *et al.*, 1985). Free sialic acids bearing these groups may also be resistant to degradation by sialic acid aldolase.

Sialidases (neuraminidases)

This large family of enzymes catalyzes the hydrolysis and release of sialic acids bound in an α-*O*-linkage in glycoproteins, glycolipids and saccharides. They are therefore α-*O*-glycosidases or α-ketosidases (Scheme 14).

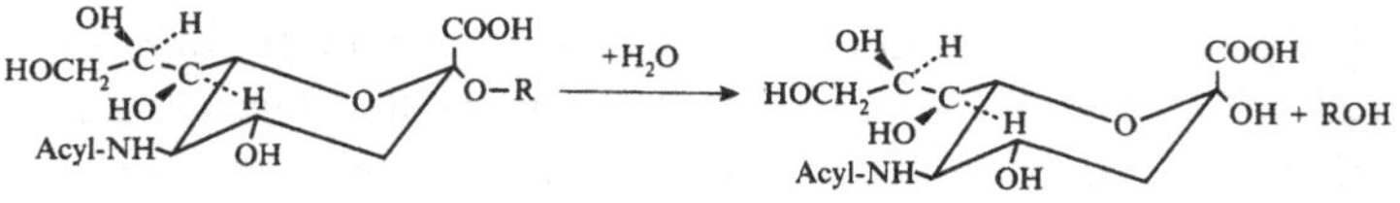

$$H_2O$$
$$R\text{-}O\ Neu5Ac \rightarrow R\text{-}OH + Neu5Ac$$

Scheme 14. Action of neuraminidase.

The study of neuraminidase has been greatly facilitated by the thiobarbituric acid assay which measures only the free form of sialic acid and is therefore ideally suited to measure the release of sialic acids from their bound state, since sialic acids bound at C-2 do not react. This sensitive assay depends upon the formation of formylpyruvic acid from carbon atoms 1–4 of sialic acid by periodate oxidation. Formylpyruvic acid combines with two molecules of thiobarbituric acid to form a chromophore with an intense red color (Warren, 1959; Aminoff, 1961).

The discovery of neuraminidase was of particular significance because it led to the elucidation of the sialic acids and the dramatic expansion of our understanding of protein and lipid-bound carbohydrates. Since most sialic acid in nature is bound in a terminal position, the enzyme has proven to be a powerful analytic tool because the functional activity of glycoproteins and glycolipids can be tested with and without their complement of sialic acid residues. The enzyme was the first to be associated with a virus, which until then was considered, by definition, to be devoid of enzymes.

Sialidase was described in culture filtrates of *Clostridia welchii* (McCrea, 1947) and *V. cholerae* (Burnet and Stone, 1947). It was called 'receptor destroying enzyme' because after red blood cells were exposed to the filtrates they could not be agglutinated by influenza virus. Both the filtrates and the influenza virus contain the enzyme which removes sialic acid from the surface of the red blood cell, as well as from a soluble glycoprotein such as ovine submaxillary gland mucin (Ada and Lind, 1961). Sialic acid is an essential component of the cellular receptor for the virus, since, once it is removed from the surface of the cell, the cell can no longer be agglutinated by fresh virus. Sialyl lactose and other sialic acid-containing substances that have an affinity for the virus inhibit agglutination. If the sialic acid is replaced on treated red blood cells by incubation with CMPNeu5Ac and a sialyl transferase, the cells become agglutinable. In more recent work, it has been found that the sialidase is present in influenza A and B viruses. Influenza C virus differs in that it bears a 'receptor destroying enzyme' but this enzyme specifically hydrolyses the 9-*O*-acetyl group of *N*-acetyl-9-*O*-acetylneuraminic acid bound in human nasal mucins (Herrier *et al.*, 1985). In early studies, the nature of the liberated molecule was unknown, but later it was shown to be sialic acid (Gottschalk and Lind, 1949). Sialidase was first described in mammals in 1960 (Warren and Spearing, 1960) and in chick embryos in 1961 (Ada and Lind, 1961).

Sialidases of virus (myxovirus and paramyxovirus), bacteria, and from many tissues of animals, have been purified and extensively studied. They operate best at an acidic pH. Within the animal cell, one form of sialidase has been found in the lysosomes (Tulsiani and Carubelli,1970), and other forms, with different properties, in the microsomes (Miyagi and Tsuiki, 1985) and plasma membranes (Schengrund, Rosenberg and Repman, 1976). All forms of the enzymes are inhibited by the deoxy-dehydrosialic acid analog 5-acetamido-2,6-anhydro-3,5-dideoxy-D-glycero-D-galacto-non-2-enonic acid (Warner *et al.*, 1991). The molecule has the general shape of sialic acid, but lacks a reducing group at C-2, so that it cannot form a link with an hydroxyl group of a second sugar.

This potent inhibitor has proven useful in distinguishing between the various sialidases, which not only have different substrate specificities but also differ in their susceptibility to the inhibitor. The deoxy-dehydro analog has been used in studies supporting the hypothesis that an extracellular sialidase can affect the growth of human fibroblasts. Inhibition of this enzyme by the analog can reduce or abolish cell growth (Usuki, Hoops and Sweeley,

1988). The epidermal growth factor (EGF) stimulates growth by combining with a receptor on the cell surface to induce an autophosphorylation of the receptor. The process, part of growth regulation, is inhibited by the ganglioside GM3, which is also a component of the cell surface membrane. An extracellular sialidase, present in tissue culture medium, can remove the sialic acid of GM3 to form ceramide lactose and free sialic acid thus eliminating the inhibition of cell division (Usuki *et al.*, 1988).

There are numerous investigations which show that sialic acids are an essential component of receptors at the cell surface for viruses, toxins, antibodies, hormones and lectins, bacteria, and other cells. In these instances, sialic acids are signals for recognition. However, they can also mask recognition sites on molecules or on cell surfaces.

Ashwell and Morell (1981) enzymatically removed sialic acid from soluble glycoproteins to expose an underlying galactose residue to which sialic acid had been bound. The asialoglycoproteins, when injected into an animal, were picked up by the liver, because liver cells bear galactose-specific receptors on their surfaces. Desialylated erythrocytes and lymphocytes are also taken up by liver and spleen. However, lymphocytes apparently resialylate their surface structures and reappear in the blood (Kaufmann, Schauer and Hahn, 1981). The removal and replacement of sialic acid at the cell surface serves as a model for control of cell interactions. It could be an important mechanism for dissemination of tumor cells and for effecting changes in relationships between cells during differentiation of specific cells in the embryo.

Removal of sialic acids may be particularly important because the overall negative charge on the molecule to which they are bound is reduced, and they may become susceptible to a stepwise degradation by a series of glycosidases as well as by proteases. The cell can render the sialic acid partially or completely resistant to removal by sialidases by acetylating one or more of the hydroxyl groups on the sialic acids. Similarly, a methyl group on the 4-hydroxyl group of bound sialic acid (Neu5Ac4Me) renders

it resistant to bacterial and viral neuraminidase, and free
Neu5Ac4Me is resistant to degradation by sialic acid aldolase
(Beau and Schauer, 1980).

In contrast to the sialidases that remove only terminal sialic acid
molecules, an endosialidase associated with a bacteriophage growing in a strain of *Esch. coli* KI, is capable of cleaving the α2-8
linkage between two residues of sialic acid in polysialosyl
sequences found in colominic acid and in neural cell adhesion
molecules (Vimr *et al.*, 1984; Finne and Mäkela, 1985). Colominic
acid is produced by *Esch. coli* KI. Sialic acid oligomers up to
seven residues long are liberated by the endosialidase. The specificity of the enzyme is such that there must be at least five sialic
acid residues on the proximal, reducing side of the cleavage site
attached to a macromolecule, and at least three residues on the dis-
tal side: the enzyme will not cleave polysialic acid containing less
than eight residues of sialic acid. As will be discussed, the interaction of polysialosyl molecules with other adhesion molecules,
enzymes and antibodies depends on the length of the molecule, i.e.
the number of sialic acid residues. Thus, sialidases and endo-
sialidases which shorten the molecule, as well as sialyl transferase
which catalyzes polymerization of Neu5Ac, could have a
regulatory function. Adhesiveness of polysialic acids may depend
on the conformation of the chain as well as the length of the molecule.

Genetic metabolic disorders involving sialic acids

Numerous examples of mechanisms controlling the metabolism of
hexosamines and sialic acid have been discussed. The absence of
control can have serious consequences. For instance, in the rare
genetic disease sialuria there is an overproduction of free sialic
acid with massive excretion in the urine (5–7 g per day) (Fontaine
et al., 1968; Seppala *et al.*, 1991). Patients may be mentally
retarded, have coarse facies, hepatosplenomegaly and metabolic
acidosis. Fibroblasts from these individuals contain very high

levels of free sialic acid in the cytoplasm. The defect in the cells appears to be a failure of the UDP-*N*-acetylglucosamine 2'-epimerase to be inhibited by CMP-sialic acid, leading to an excess formation of *N*-acetylmannosamine (Reaction 12) that is converted into sialic acid. Not only is there a critical loss of the feedback inhibitory mechanism, remarkably, there also appears to be an enhanced epimerase activity as well.

Other metabolic disorders involving the catabolism of sialic acids have been described in man. In Salla's disease, there is severe mental retardation, hypersecretion of sialic acid in the urine, and retention of free sialic acid within vacuolated lysosomes. Free sialic acid, formed by cleavage of glycoproteins and glycolipids by lysosomal neuraminidase, cannot be transported out of the lysosome where it accumulates (Renlund, Tietze and Gahl, 1986). Infantile free sialic acid storage disease suffers the same defect as in Salla's disease but is more severe and leads to early death (Mancini *et al.*, 1986). Finally, in the disease sialidosis, there is an accumulation of sialic acid-containing molecules in the lysosomes due to an absence of lysosomal neuraminidase (Beaudet, 1983). Degradation of saccharide-containing molecules in the normal course of turnover does not take place without an initial removal of terminal sialic acid.

Glycosidases

Glycosidases, the ubiquitous lysosomal enzymes, found both intra- and extracellularly, are capable of degrading lipids, proteins, nucleic acids, and sugar polymers. They play an important part in normal metabolic turnover and the maintenance of optimal levels of various molecules (hormones, carriers, etc.), and in defense against infection. The glycosidases hydrolytically and irreversibly cleave single, terminal sugars from small and large carbohydrate structures to release a free sugar. They operate best at pH values of 4.5–6.0, and several of these enzymes, working in succession, can degrade a sugar polymer in a stepwise manner. The glycosidases are specific for sugars and, since these are bound at the C-1 position in an α or β

configuration, there are two major groups of enzymes: α- and β-glycosidases. Thus, there are α- and β-galactosidases and α-and β-*N*-acetylglucosaminidases. Since fucose is bound by an α-linkage, only α-fucosidase is found and, for similar reasons, sialidase has only an α- specificity. In addition to the lysosomal glycosidases, there are also highly specific, membrane-bound glycosidases (glucosidase I and II, and mannosidase I and II) in the ER and Golgi that are responsible for the trimming of the saccharide transferred from dolichol to asparagine within the lumen of the ER. As discussed, these enzymes are blocked by quite specific inhibitors.

In addition to the glycosidases that act only on sugar in a terminal position, there are endoglycosidases such as endosialidase that hydrolyzes the bond between interior sialic acid residues of polysialic acid structures. There is also an endo-β-galactosidase acting on *N*-acetyllactosamine units of glycoproteins, polysaccharides and glycolipids. This enzyme cleaves the β1-4 linkage between galactose and GlcNAc or between galactose and glucose when they occur within a linear unbranched polymer, but not when the galactose is in a β1-3 linkage or is terminal.

$$- - - \text{gal } \beta\text{1-4 GlcNAc } \beta\text{1-3} - - -$$
$$- - - \text{gal } \beta\text{1-4 Glc } \beta\text{1-3} - - -$$

Several endo-*N*-acetylglucosaminidases have been described in animals, bacteria and moulds that cleave the β1-4 linkage between two GlcNAc residues one of which is attached to the amide N of asparagine within the polypeptide. Thus these enzymes, which can strip a glycoprotein of virtually all of its carbohydrate, are invaluable analytic tools. Endoglycosidase D acts on *N*-linked high mannose oligosaccharides from glycoproteins and glycopeptides (Mizuochi, Amano and Kobata, 1984). Endoglycosidase F cleaves *N*-linked high mannose, bi-antennary hybrid and complex *N*-linked structures (Tarentino, Gomez and Plummer, 1985), while endoglycosidase H does not act on complex structures but will cleave high mannose oligosaccharides (Tarentino and Maley, 1974).

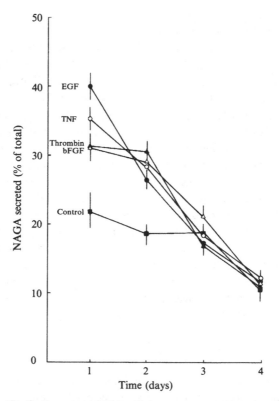

Fig. 3. Secretion of *N*-acetylglucosaminidase (NAGA) by PyS-2 cells. Cells were plated and cultured in 30-mm plastic Petri dishes, and each day thereafter duplicate cultures were assayed for NAGA. Factors were added to each dish: epidermal growth factor (EGF), 20 ng/ml; basic fibroblast factor (b-FGF), 20 ng/ml; thrombin, 0.5 units/ml; tumor necrosis factor (TNF), 100 units/ml. (Taken from Warren (1990). *Exp. Cell Res.* **190**, 133 with permission.)

Substantial glycosidic enzyme activity is present in the serum. One would expect that this would lead to extensive degradation of the carbohydrate component of serum glycoproteins. This in turn should make these structures more susceptible to proteases. Since the glycosidases themselves are glycoproteins, they could be their own substrates. Yet, there is no evidence that significant degradation takes place. One possible explanation is that terminal sialic acid on glycoproteins must first be removed before the other glycosidases can act and there is only an extremely small amount of sialidase activity in bovine and human serum (Warren and Spearing, 1960). However, we have found that removal of serum sialic acid by added bacterial sialidase does not increase the degradation of serum glycoproteins by endogenous serum glycosidases (Warren, unpublished). Perhaps there are protective, endogenous inhibitors in the serum.

It is known that the levels of extracellular lysosomal glycosidases are relatively high in tumors and especially so at their periphery (Sylven and Bois-Svensson, 1965). This is consistent with an elevation of serum glycosidases in humans and animals bearing tumors (Bernacki, Niedbala and Korytnyk, 1985). By degrading intercellular matrix, lysosomal enzymes could facilitate invasion and metastasis of tumor cells (Liotta, 1986). Using an assay for secretion of N-acetylglucosaminidase (NAGA) by cells in culture (Warren, 1989), it has been found that transformed, malignant cells have a higher rate of secretion than control cells, and that they are stimulated to a much greater extent by epidermal growth factor, basic fibroblast growth factor, transformation growth factor, thrombin and tumor necrosis factor (Fig. 3). Stimulation is greatest in the log phase of growth (Warren, 1990). These observations suggest that altered or increased responsiveness to paracrine and autocrine growth factors *in vivo* may be responsible for both persistent cell division and increased secretion of degradative enzymes: hallmarks of the invasive malignant cell.

Bound carbohydrates in malignancy

In recent years, cancer research has been dominated by the oncogene theory that was brought to light and elaborated by the astonishing power of molecular biology. Whatever the new insights provided by this approach in explaining the origin and development of malignancy, the problem of how cancer cells express their malignancy remains. There is good reason to believe that the surface of the cell is involved in an important way, although it is unlikely that changes at this site are a cause of malignancy. There are convincing data that the cell surface structure is involved in changes in intercellular adhesiveness, cytoskeletal structure, transport of metabolites and ions, binding of various growth factors and cell behavior modifiers, and in immunological characteristics. Changes in one or more of these functions could be essential for the expression of malignancy.

The bound carbohydrates are concentrated at the surface of the cell. At least 66% of the cell's sugars are in or on the surface membrane and almost all face the outside of the cell. Does this important surface constituent change in the malignant cell and, if so, what processes are affected? How is the change brought about? These questions have been examined since the mid-1970s and the data have been reviewed (Warren, Buck and Tuszynski, 1978; Smets and van Beek, 1984; Dennis, 1988; Kobata, 1988). Clearly, there are changes in the bound carbohydrates of the malignant cell and it remains to be seen whether they can be used for diagnosis, prognosis or immunotherapy.

A major change that takes place in the glycoproteins of malignant cells is an enlargement of their carbohydrate groups. This extensive change takes place in 80% of the glycoproteins of the cell (Tuszynski *et al.*, 1978). As discussed earlier, altered structure may lead to altered function however 'minor' so that numerous, small changes in function could occur associated with every membrane system of the cell, including the nuclear membrane. Although not lethal, they might affect intercellular relations and the fine control of mitosis.

The enlargement of the glycoprotein carbohydrate groups can be demonstrated by the double-label method previously discussed. Fig. 4 shows the profiles of ^{14}C and ^{3}H-labeled glycopeptides derived from control (BHK_{21}/C_{13}) and Rous virus-transformed

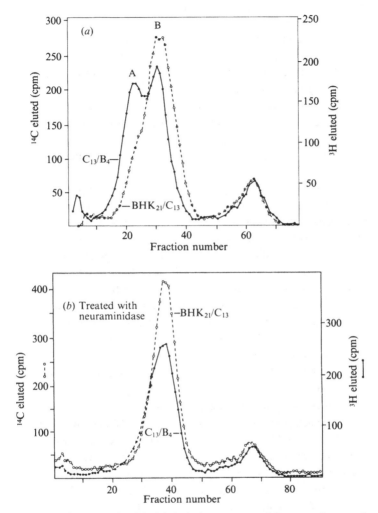

Fig. 4. (a) Double-label elution patterns of Pronase digests of glycoproteins from control BHK_{21}/C_{13} hamster cells, [^{14}C]fucose (○) and from Rous virus transformed cells C_{13}/B_4, [^3H]fucose (●). Glycopeptides were eluted from a column of Sephadex G50. The earlier eluting, peak A glycopeptides are larger (fractions 15–27). (b) The Pronase digests were treated with 5 units of neuraminidase (*V. cholerae*) for 1.5 h at pH 5.2 in 3 mM $CaCl_2$, sodium acetate, before application to the column.

cells C_{13}/B_4 , and fractionated on the basis of size, on a long column of Sephadex G50. The glycopeptide consists of the intact carbohydrate group bearing only a few amino acids and was derived from the glycoprotein by exhaustive digestion by pronase. It is apparent that there is an enrichment of a population of glycopeptides (peak A), which is significantly increased in size in material from the transformed, malignant cell.

Increased amounts of peak A glycopeptides are found in chick embryo fibroblasts transformed with Rous sarcoma virus, but not in untransformed cells or cells infected only with Rous associated (helper) virus. When chick embryo fibroblasts are transformed with temperature-sensitive mutants of Rous sarcoma virus, the cells, grown at permissive temperature (36 °C) where transformation is expressed, overproduce peak A carbohydrates; they do not do so at the nonpermissive temperature of 41 °C (Warren, Critchley and MacPherson, 1972a). Enhanced amounts of peak A glycopeptides are found in various solid murine tumors (lymphosarcoma, melanoma, mammary carcinoma) compared to corresponding normal tissues, and in 3T3 mouse fibroblasts transformed by SV40, polyoma and mouse sarcoma virus. Virus-transformed cells overproduce peak A glycopeptides whether or not the infection is productive of new virus. Increased peak A glycopeptides are also found in ascitic and solid hepatomas, in chemically induced malignancies, and in hamster cells transformed by DNA and RNA oncogenic viruses and by chemicals. In the human, the increase of peak A glycopeptides is found in acute and chronic lymphatic and myeloid leukemia, as well as in Burkitt's lymphoma, breast carcinoma, and neuroblastoma (Smets and van Beek, 1984). The glycoproteins of circulating lymphocytes, lymphoblasts from patients with infectious mononucleosis, and phytohemagglutinin-stimulated lymphocytes, have normal complements of peak A glycopeptides (van Beek, Smets and Emmelot, 1975).

In summary, the phenomenon occurs almost without exception in five species of cells including fibroblasts, epithelial, blood, and liver cells rendered malignant by viruses or chemicals, or arising spontaneously. The overexpression of peak A glycopeptides may be more closely associated with *in vivo* tumorigenicity than with the state of transformation in culture. Hamster fibroblasts transformed with the chemical carcinogen dimethylnitrosamine produce the same amount of peak A glycopeptides as control cells. Yet, they grow in soft agar, continue to divide even after contact, and look transformed. However, when these cells are injected into hamsters and subsequently re-isolated from tumor masses for growth and double-label comparisons in culture, there is a marked increase in peak A glycopeptides on their cell surfaces. Several lines of isolated cells have been tested for their latent period of induction after re-injection into hamsters. As the latent period declines, the amount of peak A saccharide on their surface membrane glycoproteins increases. Variant cells with a lower tumorigenicity have been re-isolated and the glycopeptides from their cell surfaces have been found to contain less group A material (Glick, Rabinowitz and Sachs, 1973). Similar results have been found using a hamster cell transfected with polyoma virus (Glick, Rabinowitz and Sachs, 1974) and SV40-transfected 3T3 cells (Smets and van Beek, 1984).

Changes in the levels of peak A glycopeptides in cells could be due to shifts in the population sizes of individual glycoproteins. Is the increase due to overexpression of glycoproteins that happen to be enriched in peak A carbohydrate groups or is it due to the enlargement of the carbohydrate groups on polypeptides? To resolve the problem, control BHK_{21}/C_{13} cells and transformed malignant C_{13}/B_4 cells were labeled with [^{14}C] and [^{3}H]glucosamine, respectively. Individual glycoproteins (24) were isolated from each cell by extraction and by purification by SDS-polyacrylamide gel electrophoresis and isoelectric focusing on 5% polyacrylamide slab gels. Homologous glycoprotein bands were cut out of the gel, and comparable ^{14}C and ^{3}H-labeled glycoproteins were mixed and exhaustively digested with Pronase. The resulting mixture of [^{14}C] and [^{3}H]glycopeptides were analyzed on columns of Sephadex G50. The elution patterns of only four of

24 glycoproteins were the same. Their M_r ranged from 31 000 to 179 000 and they had come from all membrane systems of the cell. These data tell us that in malignant cells the glycoproteins bear more, larger carbohydrate groups (Tuszynski *et al.*, 1978).

Since malignant cells persist in dividing, the overproduction of peak A glycopeptides could be associated with mitosis and the growth state – an exaggeration of a normal state. In fact, dividing, non-malignant hamster cells (Buck *et al.*, 1971) and stimulated human lymphocytes (van Beek *et al.*, 1975) are enriched in peak A glyco-peptides. Peak A glycopeptides are more prominent in glycoprotein of normal hamster cells in metaphase (Glick and Buck, 1973).

The relative amounts of glycopeptides of chick chondroblasts growing in monolayer, in suspension culture, as a pellet and in carti-lage itself, have been compared. As seen in Fig. 1 (p. 14), increased synthesis of larger protein-bound carbohydrate groups occurs in monolayer and suspension culture, where close contact and restraints on growth are removed compared to that in cells in carti-lage itself, or in a compacted state in a pellet (Cossu *et al.*, 1982b).

Peak A glycopeptides from transformed cells have been treated with neuraminidase and then compared with an untreated standard on a column of Sephadex G50. From Fig. 4b it is clear that the gly-copeptides have been reduced in size so that they co-chromatograph with type B glycopeptides. However, peak A and B glycopeptides differ in ways other than their sialic acid content. Evidently, peak A glycopeptides are particularly rich in sialic acid and, as will be dis-cussed, contain extra antennae capped by a sialic acid residue.

A considerable amount of work has been done on elucidating the essential structure of the peak A glycopeptides (Warren *et al.*, 1978; Kobata, 1988; Dennis, 1988) as compared to smaller, peak B glycopeptides. While the latter are bi-antennary, peak A gly-copeptides are larger tri- and tetra-antennary structures, each with a terminal sialic acid group (Yamashita *et al.*, 1984). The increase in antennae appears to be due to an enhanced *N*-acetylglu-cosaminyl transferase V which transfers GlcNAc from UDPGlcNAc to a mannose residue in a β1-6 linkage. The antenna is completed by the activity of a galactosyl and a sialyl transferase. This has been found in hamster cells transformed by polyoma virus (Yamashita *et al.*, 1985) and Rous sarcoma virus (Arango *et al.*, 1987), as well as in breast carcinoma (Dennis and Laferte, 1989). The level of the enzyme is also high in normal intestinal epithelial cells, which, like malignant cells, are motile and rapidly dividing (Dennis *et al.*, 1989).

Tumorigenicity and metastatic potential are associated with decreased adhesiveness and increased cell motility, which are, possibly, dependent on increased levels of peak A carbohydrate groups. The larger size of the glycopeptides, due to increased α1-6 branching, could cause steric hindrances that decrease adhesiveness between molecule and molecule, or molecule and cell. In addition to an increase in size, the carbohydrate groups carry more negatively charged sialic acids that could also interfere with adhesiveness. There are several examples of increased adhesiveness of cells to substrate upon removal of sialic acid by sialidase. As the terminal galactose and GalNAc residues on glycoproteins are substituted with sialic acid, and as the number of antennae terminated by a sialic acid molecule increases, the tumorigenicity and metastatic properties increase and the immunological characteristics of the cell surface probably change.

A growth-dependent sialyl transferase has been described that increases at least 2.5–11-fold in amount in virus-transformed hamster cells compared to non-transformed controls. The acceptor is a desialylated glycopeptide from transformed cells. The sialyl transferase is apparently specific for this acceptor because similar sialyl transferase activities have not been found in control and transformed cells when endogeneous activity is measured, or desialylated fetuin is used as an acceptor (Warren, Fuhrer and Buck, 1972b). Enhanced sialyl transferase activity (1.5–6-fold), using acceptors prepared from transformed cells, has been found in polyoma murine sarcoma- and Rous sarcoma virus-transformed 3T3 cells (Bosmann, 1972). However, this change in sialyl transferase activity is not found in all transformed cell systems.

A decrease in adhesiveness, so important in malignancy, has been ascribed to altered cell surface glycoproteins. However, as previously discussed, there is extensive enlargement of the carbohydrate groups throughout the cell with, presumably, attendant functional changes. The consequences are unknown, but there may be numerous glycoprotein-associated or glycoprotein-dependent processes that are affected. The shifts are apparently not lethal, but out of the scramble may arise pleiotropic, malignant dysfunction, especially in subtle activities such as cell behavior.

Enlargement of carbohydrate groups, with an increase in charge, that contribute to the malignant state may also arise in other ways. As previously discussed, glycoproteins bearing polysialic acids as well as polylactosamine groups are present in embryonic cells. The rates of synthesis of these carbohydrate groups are programed during embryogenesis where orderly changes in synthesis are followed by desired changes in intercellular adhesion, migration and rates of cell division. As synthesis of the bound polysialic acid and polylactosamine comes to a halt, a programed intercellular adhesiveness commences. There are several examples of adhesiveness increasing upon treatment of cells with endo-galactosidase which degrades polylactosamine. In the malignant cell, control of gene expression is defective and there may be disorderly and inappropriate re-expression of these molecules, resulting in a decreased intercellular adhesiveness and other malignant characteristics in adult cells (Hakomori and Kannagi, 1983).

Increase in β1-6 branching due to increased N-acetylglucosaminyl transferase V activity may be associated with the re-expression of embryonic polysialic acid and polylactosamine on glycoproteins, for there is some evidence that synthesis of these molecules in the cancer cell tends to take place on new 'extra' β1-6 GlcNAc residues (Dennis *et al.*, 1989). Normal expression of these molecules has been discussed previously (p. 45). Polysialic acid associated- and tri- and tetra-antennary asparagine-linked carbohydrate groups are re-expressed on glycoproteins in Wilm's tumor, especially in regions of the tumor that do not adhere to substratum (Roth *et al.*, 1988). Bound polylactosamine structures are found in thyroid carcinomas but not in benign thyroid cells (Yamamoto *et al.*, 1984). On the other hand, polylactosamine structures associated with lysosomal membrane glycoproteins decrease as human colonic adenocarcinoma cells in culture differentiate spontaneously (Youakim *et al.*, 1989), just as polylactosamine on fibronectin disappears when teratocarcinoma cells (Cossu and Warren, 1983) or promyelocytic leukemic HL60 cells are induced to differentiate (Cossu *et al.*, 1982a).

Transfer of GlcNAc to the 3'-hydroxyl of galactose during polylactosamine synthesis is catalyzed by a β1-3 GlcNAc transferase that is unexpressed in differentiated cells (Phillips *et al.*, 1990), while β1-4 galactosyl transferase activity is wide-spread and in relatively high concentration. The programed appearance or disappearance of polylactosamine structures is apparently controlled by β1-3 GlcNAc transferase activity as well as by the rate of production of acceptor tri- and tetra-antennary structures.

There is other support for the notion that enlarged, charged, carbohydrate groups on surface glycoproteins are responsible for decreased adhesiveness and lead to a greater malignant potential. In general, mutants of transformed cells that are unable to synthesize large groups lose their metastatic potential and revertants regain the property (Dennis *et al.*, 1989). If synthesis of larger tri- and tetra-antennary groups are inhibited by swainsonine, metastatic potential and invasiveness are decreased, while adhesion of tumors to basement membrane is increased (Dennis *et al.*, 1989). At the molecular level, it has been shown that stripping sialic acid and polylactosamine from a lysosomal membrane-associated glycoprotein (LAMP-1) by sialidase and endo-galactosidase respectively, increases its affinity for extracellular matrix proteins (Dennis *et al.*, 1989). It will be of the utmost interest to insert the gene encoding GlcNAc transferase V into a non-malignant cell to see whether it takes on the character of an immortalized, transformed cell or whether increasing expression of this transferase in a malignant cell by transfection will enhance its tumorigenicity.

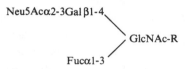

Scheme 15. Sialyl-Le^x antigen.

Other carbohydrate groups are also involved in intercellular adhesion. Recently, it has been shown that adhesion of myeloid cells to the endothelial cells of blood vessel walls is mediated by a carbohydrate ligand, sialyl-Lewis X (Scheme 15), found in terminal positions on glycoproteins and glycolipids of neutrophils.

The endothelial leukocyte adhesion molecule 1 (ELAM-1) is induced during infection or when vascular endothelial cells are exposed to interleukin-1 and tumor necrosis factor. The molecule recognizes the sialyl-Le^x group on the neutrophil, and adhesion follows. Antibodies to this group inhibit adhesion between neutrophil and endothelial cells. Anti-sialyl-Le^x monoclonal antibody also inhibit binding of soluble ELAM-1 to myeloid cells (Benkovic *et al.*, 1990; Phillips *et al.*, 1990). The union of cells bearing the sialyl-Le^x group with ELAM-1 on endothelial cells is the first step in the passage of the cell from the blood vessel into tissue. Although this takes place in the inflammatory process, it bears great similarity to an essential step in the metastatic spread of tumor cells.

It should be stressed that in many of the studies discussed, which are only a small fraction of the total, there is relatively little rigorous proof that a specific change in cell behavior is due to a specific change in carbohydrate structure: one only claims an association. The processes studied are poorly understood for they are extremely complex and their underlying mechanisms are essentially unknown. Further, there are few studies in which changes in the total amount of a glycoprotein in a given cell are considered along with the qualitative changes in the bound carbohydrate.

3

Multidrug resistance

A major problem in cancer chemotherapy is the development of resistance by malignant cells to drugs such as vinca alkaloids or anthracyclines. Resistance may be inherent or develop after exposure to chemotherapeutic agents.

There are several mechanisms by which a cell develops resistance. My colleagues and I are concerned with a major form of multidrug resistance (MDR) that is associated with the presence of a glycoprotein of M_r 160 000–180 000 called P-glycoprotein (P-170) in the surface membrane of the resistant cell. In this form of drug resistance, a cell that has been exposed to a single drug expresses P-170 on its surface and simultaneously becomes resistant to a variety of other drugs of different structures that may operate by different mechanisms (Beck, 1987; Gottesman and Pastan, 1988; Endicott and Ling, 1989; van der Bliek and Borst, 1989; Burt *et al.*, 1990; Ford and Hait, 1990).

The drugs to which the cell is resistant are usually lipophilic, with an N-containing ring and they have a slightly positive charge at physiological pH (Zamora, Pearce and Beck, 1988). Drug resistance can be at least partially reversed by less toxic agents such as verapamil (Tsuruo *et al.*, 1981) and quinidine (Fojo *et al.*, 1987) that inhibit the transport function of P-170. There is much evidence to show that P-170 is a relatively non-specific, energy-requiring pump with an affinity for ATP, and it is widely believed that drugs of a lipophilic nature, entering the cell, are pumped out rendering the cell resistant (Horio, Gottesman and Pastan, 1988).

In this chapter, I will argue that P-170 in the MDR cell not only increases efflux but also causes a decreased influx of drug and that the contribution of each to drug resistance depends on a balance of

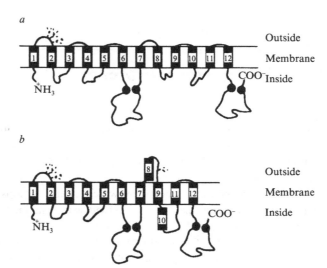

Fig. 5. (*a*) Conventional model of P-glycoprotein with 12 transmembrane domains (thick lines) and carbohydrates bound to
the first extracellular loop. (*b*) New topological structure model
proposed with a different orientation in the arrangement of
hydrophobic domains 8, 9 and 10 based on recent data (Zhang
and Ling, 1991). The solid circles are the ATP-binding sites.
Note the additional glycosylation site between hydrophobic
segments 8 and 9 in (*b*). (Figure taken from Zhang and Ling,
1991.)

the two. The balance may vary with the type of cell and perhaps its
condition and age, as well as with the particular drug involved.
The work described was carried out in collaboration with Dr Jean-
Claude Jardillier of the Institute Jean Godinot, Reims, France.

Properties of P-glycoprotein

P-glycoprotein is a tandemly duplicated molecule, 1280 amino
acids long (Fig. 5) that has 12, α-helical, transmembrane segments
with six extracellular loops and a central loop or linker region. Of
the first half of P-170 43% is the same as the second half suggest-

ing that P-170 evolved by a gene duplication. Each half of the molecule contains an ATP-binding site which is on the cytoplasmic side of the plasma membrane, and the hydrophobic, α-helical, membrane-spanning domains are arranged so that they probably form a pore through which a variety of lipophilic drugs are expelled from the cell. In some unknown manner, the hydrolysis of ATP bound to the ATP-binding domains is linked to the expulsion of the drug. The asparagine N-linked carbohydrate of P-170, which is on the first extracellular loop, has an M_r of approximately 30 000. Recent work has revealed that an additional extracellular carbohydrate group may be present on the loop between the eighth and ninth hydrophobic domains and the arrangements of the transmembrane segments 8, 9 and 10 may differ from that previously accepted (Fig. 5*b*) (Zhang and Ling, 1991). In this scheme there are only 10 transmembrane segments, compared to the usually accepted 12 spanners (Fig. 5*a*). Perturbation of the structure of the carbohydrate groups has no major effect on the function of P-170, nor on MDR. This has been shown by growing MDR cells in the presence of tunicamycin, by treating intact MDR cells with pronase to remove surface peptides bearing carbohydrates, and by examining mutant MDR cells which are defective in carbohydrate pathways. In all instances, the rate of accumulation of drugs and drug toxicity was the same as in MDR cells with unaltered P-170 (Endicott and Ling, 1989; Juranka, Zastawny and Ling, 1989).

As shown by photoaffinity labeling, P-170 binds a drug such as a vinblastine analog. Binding can be abolished by vincristine or daunorubicin as well as by an MDR-reversing agent such as verapamil or quinidine. Similarly, attachment of a photosensitive analog of verapamil can be inhibited by vinblastine which indicates that they are all competing for a common site (Beck, 1987; Gottesman and Pastan, 1988).

The genes coding for P-170 have been cloned. Two genes, MDR1 and MDR2, have been identified in the human genome, but only the expression of MDR1 renders a cell drug-resistant. In other species, there may be several homologous MDR genes; there are three in both the hamster and mouse. Mouse and human P-170 are 80% identical in sequence (Juranka *et al.*, 1989).

P-170, a highly conserved structure, has been found in normal tissues, in epithelial cells of the liver, pancreas, kidney, small intestine and colon, as well as in the adrenal cortex. It is believed that P-170 in these organs, which is found on the luminal surface of epithelial cells of gut or ducts, functions to rid cells and the organism of noxious agents. In the adrenal, the molecule may be involved in the transport of steroids. It is probably no coincidence that tumors arising from tissues that normally express P-170 are particularly resistant to chemotherapy (Juranka *et al.*, 1989).

Changes in the chromosomes of drug-resistant tumor cells are sometimes found that suggest increased genomic material due to gene amplification leading to overexpression of P-170: homogeneous staining regions and/or double minutes are observed (Burt *et al.*, 1990). However, there may be overexpression of P-170 in the absence of gene amplification, probably due to deregulation of transcription of P-170 mRNA.

There is considerable homology in the amino acid sequences of P-170 and a protein STE 6, found in yeast, which is necessary for the secretion of a mating pheromone (α factor). Yeast lacking STE 6 function normally but are sterile (van der Bliek and Borst, 1989). Genes coding for structures at least partially homologous with those of P-170 have been detected by hybridization to molecular probes in *Drosophila* (Juranka *et al.*, 1989) and *Plasmodium falciparum* (Martin, Oduola and Milhous, 1987). Like the MDR genes, the invertebrate genes are tandemly duplicated. In plasmodium, the gene is amplified in chloroquine-resistant cells (Wilson *et al.*, 1989; Foote *et al.*, 1989). Just as in MDR animal cells, efflux of drug from plasmodium is increased and drug resistance can be reversed with verapamil (Krogstad *et al.*, 1987). Thus, members of a multigene family coding for transport (glyco)proteins play a crucial role in the therapy of two of the major diseases of mankind, cancer and malaria. Another important member of this family is the cystic fibrosis transmembrane conductance regulator, probably responsible for transmembrane anion transport and defective in cystic fibrosis (Riordan *et al.*, 1989). Bacterial permease system components are also members of this family of transporters (Juranka *et al.*, 1989).

Experimentally, MDR has been studied in cells in culture. MDR cells have been generated by challenging cells with drugs such as vinblastine, colchicine, vincristine, doxorubicin or daunorubicin, and then selecting survivors. These are again challenged with increased amounts of drugs. Although the MDR cell lines developed by this approach have proven to be useful, unknown but significant changes other than drug-resistance may also occur. A more specific and satisfactory method is the transfection of cells with plasmids containing the MDR1 gene (Ueda *et al.*, 1987). These experiments have shown that full multidrug resistance can be conferred by a single gene coding for P-170 and that other glycoproteins that have been occasionally described in MDR cells are not essential. Further, transfection of cells with specifically mutagenized MDR1 permits the functional analysis of various parts of the P-170 molecule. For instance, a point mutation that results in substitution of a valine instead of glycine at position 185 results in an altered pattern of drug resistance (preferential resistance to colchicine) (Choi *et al.*, 1988).

We have worked mainly with a human lymphoblastic leukemic cell line CEM, and its MDR counterpart CEM/VLB$_{100}$ which is capable of full growth in medium containing vinblastine (VLB) at a concentration of 0.4 µg/ml. In addition, we have used other pairs of drug-sensitive and P-170-containing MDR cells, KB-3-1 and KB-V1 (human lung carcinoma), K562 and K562 ADM (human erythroleukemia), NIH3T3 and MDR3T3 (mouse fibroblast) and its counterpart transfected with the human MDR1 gene (Ueda *et al.*, 1987).

Lysosomes and multidrug resistance

It has been postulated that lysosomes and other intracellular, acidified vesicles are important components of the drug-resistance

Table 4. *Lysosomal enzyme activity of cells sensitive and resistant to drugs*

Cell	Total NAGA[a]	Total β-galactosidase[a]	Protein (μg/10^6 cells)
CEM	143.8 ± 17.8[b] (16)[c]	49.7 ± 6.3 (3)	59.7 ± 7.7 (16)
CEM/VLB$_{100}$	32.9 ± 5.3 (16)	9.4 ± 2.0 (3)	59.8 ± 4.9 (16)
CEM	145.7 ± 15.6		
CEM/VM-1	168.4 ± 18.1		

Notes: CEM is the human lymphoblastic leukemic cell and CEM/VLB$_{100}$ its P-170-enriched, MDR counterpart. CEM/VM-1 has MDR by a non-P-170 mechanism. NAGA, *N*-acetylglucosaminidase.
[a]Substrate hydrolyzed (nmol/10^6 cells per 30 min); [b]mean ± SEM; [c]number of experiments.
Source: Enzyme activity described in Warren *et al.*, 1991.

mechanism because they can sequester basic, cytotoxic drugs such as vinblastine or doxorubicin, removing them from sensitive, target sites. Drugs can either be rendered inactive or they can be secreted and the role of P-170 is, in some unknown manner, to enhance secretion. Consistent with this notion is the observation that turnover of membrane components is enhanced in MDR cells with overexpressed P-170 (Beck, 1987; Sehested *et al.*, 1987; Klohs and Steinkampf, 1988). Furthermore, it has been known for many years that the content of lysosomal enzymes of MDR cells is significantly lower than that of the parental drug-sensitive cell (Bosmann and Kessel, 1970), which could result from chronic hypersecretion of lysosomal enzymes.

We have compared the levels of *N*-acetylglucosaminidase and β-galactosidase in the human lymphoblastic leukemic cell CEM, and its P-170 enriched, MDR counterpart CEM/VLB$_{100}$ (Warren, Jardillier and Ordentlich, 1991). As seen in Table 4, the level of these enzymes is low in the MDR cell confirming the observations of Bosmann and Kessel (1970). However, CEM/VM-1, a related cell, whose drug resistance is due to an altered topoisomerase II

Table 5. *Secretion of enzymes by cells sensitive or resistant to drugs*

Cell	NAGA	β-Galactosidase
CEM	$3.8 \pm 1.6\ (7)^a$	$5.7 \pm 0.6\ (3)$
CEM/VLB$_{100}$	$14.4 \pm 2.3\ (7)$	$11.1 \pm 0.9\ (3)$
CEM	$6.8 \pm 1.5\ (6)$	
CEM/VM-1	$7.6 \pm 1.4\ (6)$	

Notes: Cell types as in Table 4, *N*-acetylglucosaminidase (NAGA) was measured by the *p*-nitrophenyl assay, and β-galactosidase by the umbelliferyl assay. Results are expressed as the percentage of total enzyme secreted in 30 min when cells are incubated in Tris, MgCl$_2$ sucrose, 150 mM NaCl.
[a]Number of experiments.
Source: Warren *et al.*, 1991.

and not to overexpressed P-170, has a normal complement of lysosomal enzymes (Warren *et al.*, 1991).

The rates of secretion of *N*-acetylglucosaminidase and β-galactosidase of CEM, CEM/VLB$_{100}$ and CEM/VM-1 cells have been studied using a method in which secretion is stimulated by NaCl (Warren, 1989). The rate of secretion of lysosomal enzymes from CEM/VLB$_{100}$ cells is significantly greater than that from CEM or CEM/VM-1 cells (Table 5). These data would suggest that the presence of P-170 in the plasma membrane leads to a chronic hypersecretion of secretable materials such as sequestered drugs and lysosomal enzymes, and an increased turnover of plasma membrane.

Despite the attractiveness of this picture, there are data that indicate that a mechanism that depends on lysosomal secretion is unlikely or is, at best, marginal. Although some agents that inhibit secretion enhance cytotoxicity of drugs, there are many that have no effect on cytotoxicity (Zamora *et al.*, 1988). We have measured the level of *N*-acetylglucosaminidase and the rate of secretion in parental NIH3T3 cells and these cells rendered drug-resistant by transfection with the human MDR1 gene (3T3MDR) and a mutant

Table 6. *Secretion of N-acetylglucosaminidase by cells sensitive and resistant to drugs*

Cell	Total NAGA[a]	% of total enzymes secreted[b]
NIH3T3	0.300 ± 0.036	12.0 ± 1.1
3T3MDR	0.088 ± 0.014	27.8 ± 6.1
pHa MDRCΔ23	0.269 ± 0.043	13.6 ± 2.4

Notes: [a]*N*-acetylglucosaminidase (NAGA) measured by substrate hydrolyzed (nmol/mg cell protein per 30 min).
[b]Results are expressed as the percentage of total enzyme secreted in 30 min when cells are incubated in TMSucrose,150 mM NaCl; mean ± SEM. Cells kindly supplied by I. Pastan and M.M. Gottesman.
Source: Methods described in Warren, 1989; Warren *et al.*,1991.

form of the gene. The cell was transfected with the MDR1 gene missing nucleotides coding for 23 amino acids at the carboxyl end of P-170 (Currier *et al.*, 1989) which extends into the cytoplasm (pHa MDRCΔ23) (Table 6).

Cells bearing P-170 with a truncated carboxyl end remain drug-resistant, for they are able to grow in the presence of 0.4 µg/ml vinblastine, but, as shown in Table 6, they secrete *N*-acetylglu-cosaminidase at the same rate as parental NIH3T3 cells and have approximately the same total NAGA content. Thus, altered lyso-somal enzyme secretion and enzyme content are not essential components of drug resistance. The results suggest that the car-boxyl terminal polypeptide of P-170 extending from the cell surface into the cytoplasm, interacts with lysosomes and other vesicles, perhaps in some meaningful physiological manner.

The greatly decreased accumulation of anticancer drugs by MDR cells with overexpressed P-170 is well documented. The almost unquestioned view is held that drugs such as vinblastine and anthracyclins enter the cell by a non-specific diffusion process and that P-170, an energy-requiring efflux pump in resistant cells, rids the cells of drugs. However, I will present evidence support-ing the view that in some unknown way, the presence of P-170 glycoprotein in the plasma membrane may reduce influx of drugs as well as expel drugs. The remainder of this chapter is a brief summary of work done with Jean-Claude Jardillier of the Institut Jean Godinot, Reims, France.

Kinetics of influx and efflux of drugs

In Fig. 6, it can be seen that the rate of accumulation of [^{3}H] vinblastine ([^{3}H]VLB) by CEM and KB-3-1 cells is far more rapid than by the corresponding MDR cells, CEM/VLB$_{100}$ and KB-3-1 cells and this has been shown even after only 20 s of intake.

By observing uptake of [^{3}H]VLB at different concentrations of drug, it has been shown that CEM cells can concentrate [^{3}H]VLB approximately 50-fold over the concentration in the medium while drug-resistant CEM/VLB$_{100}$ cells concentrate [^{3}H]VLB only 2–3-fold. Some research has shown that cytoplasmic components of drug-sensitive cells can bind drugs to a greater extent than those of P-170-containing MDR cells. However, we have found that the difference in binding is not great and occurs only in some cells.

The effect of temperature on the rate of accumulation of [^{3}H]VLB differs in drug-sensitive and MDR cells (Table 7). While the Q_{10} is approximately 2 in the sensitive cell, that of the MDR cell is 1.2. When cells are exposed to verapamil or nigericin which increase accumulation of [^{3}H]VLB by MDR cells, the Q_{10} is raised to levels approaching that of the sensitive cells. These agents, which are called chemosensitizers or modulators and reverse MDR, have little or no effect on drug-sensitive cells. The results are surprising because if resistance is due to efflux of drugs by P-170, surely a temperature-sensitive process, the Q_{10} of the low level of accumulation (influx) of drugs in the MDR cell should be 2 but in fact is only 1.2, i.e. it is virtually temperature-insensitive. When accumulation is raised by verapamil, known to inhibit efflux carried out by P-170, the Q_{10} is raised to 2. On the other hand, entry of drug into a cell is believed to take place through a diffusion process which does not require energy. The Q_{10} should be

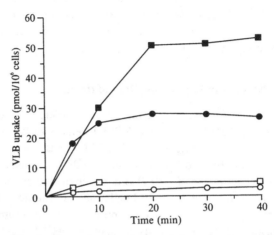

Fig. 6. Rate of accumulation of drugs by drug-sensitive and MDR cells: influx of [³H]VLB as a function of time. In these experiments, 4×10^6 CEM or CEM/VLB$_{100}$ cells in 1 ml TMSucrose solution (Tris HCl, 10 mM; MgCl$_2$, 1 mM; sucrose, 0.25 M, pH 7.1), were incubated at 37 °C with 0.5 μCi of [³H]VLB and 1.07 nmoles (1 μg) of non-radioactive VLB in a glass test tube. At various intervals, 100 μl of the cell suspension were pipetted into plastic tubes containing 6% perchloric acid solution overlayered with silicone fluid ST1250 containing 5% chloroform. The capped tubes were immediately centrifuged at 13 000 g for 20 s, and then frozen in an ethanol/dry ice bath. The tips of the tubes containing the cell pellets were cut, immersed in scintillation fluid, and their radioactivity was determined. Assays were done in duplicate, and controls in quadruplicate.

In a second set of influx experiments with 48 h cultures of drug-sensitive and MDR KB cells (human lung carcinoma) in 24-well plates, the medium in each well was removed by suction, and the cells were washed once with cold TMSucrose and then incubated at 37 °C in 150 μl TMSucrose containing 0.1 μg [³H]VLB (0.05 μCi). After incubation, the medium was removed by suction, and the cells were washed with ice-cold TMSucrose, which was removed within 7 s. Cellular radioactivity was transferred from each well to vials by two, 250 μl washes of 0.1 M NaOH and radioactivity was determined. Each

Table 7. $\mathcal{Q}_{10}$ *Values for influx and efflux of* $(^3H)VLB$

(^{3}H)VLB movement	CEM	CEM/VLB$_{100}$	KB-3-1	KB-VI
Influx	2.21 ± 0.23 (5)[a]	1.18 ± 0.13 (5)	1.98 ± 0.10 (4)	1.14 ± 0.11 (4)
+ Nigericin	2.46 ± 0.30 (2)	2.03 ± 0.31 (3)	1.89 ± 0.16 (3)	1.52 ± 0.12 (3)
+ Verapamil	1.97 ± 0.13 (2)	2.17 ± 0.19 (3)	1.83 ± 0.14 (3)	1.77 ± 0.18 (3)
Efflux	1.63 ± 0.18 (6)	1.50 ± 0.19 (6)	1.77 ± 0.30 (3)	1.70 ± 0.21 (3)

Notes: Influx and efflux experiments were carried out as described by Warren *et al.* (1991). Influx was measured after incubation for 3 and 6 min, while efflux was measured for 4 and 8 min. Incubations were at 27 and 37 °C. The $\mathcal{Q}_{10}$ is the ratio of the rates of the process at $(T+10)/T$. Concentration of drugs: nigericin, 10 μM; verapamil, 50 μM.
[a]Number in parentheses is the number of determinations, each in duplicate or triplicate.

approximately 1. We find a $\mathcal{Q}_{10}$ of less than 2 in drug-sensitive cells which suggests some special temperature-sensitive process taking place that is eliminated in the P-170-containing MDR cell where the $\mathcal{Q}_{10}$ is 1 and, as stated above, efflux pumping should be taking place. This puzzling result remains unexplained.

Direct measurement of efflux of [^{3}H]VLB from CEM and CEM/VLB$_{100}$ cells shows that the drug is lost more rapidly from drug-resistant cells (Fig. 7). During the first 10 min, CEM and CEM/VLB$_{100}$ cells lost 47.5% ± 4.9 and 63.9% ± 6.0 of their [^{3}H]VLB respectively (n=10). Thereafter, loss of [^{3}H]VLB continued slowly from CEM and more rapidly from CEM/VLB$_{100}$ cells. It should be noted that the actual amount of VLB lost from drug-sensitive cells (without P-170) during the first 10 min of efflux was

Caption for Fig. 6 (*cont.*)

incubation was done in triplicate and the results are expressed as pmol of vinblastine taken up by 10^6 cells.

Each point is the mean of values obtained in three experiments. SEM < 10%. Drug-sensitive cells: CEM, (●); KB-3-1, (■); drug-resistant cells: CEM/VLB$_{100}$, (○); KB-VI (□).

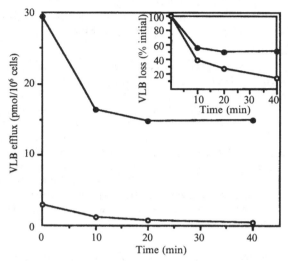

Fig. 7. Efflux of (^{3}H)VLB from drug-sensitive and drug-resistant cells. Cells were loaded with (^{3}H)VLB for 30 min, washed once with ice-cold TMSucrose, and incubated in TMSucrose containing 5 mM glucose (Warren *et al.*, 1991). Radioactivity in cells was determined in duplicate before and at various times during incubation. CEM (●) and CEM/VLB$_{100}$ (○). SEM, <13.5% ($n = 3$). Inset shows efflux as a percentage of the initial load lost with time.

far greater than that lost from MDR cells. This holds for other drug-sensitive and MDR pairs such as the human lung carcinoma cells KB-3-1 and KB-VI.

At the commencement of efflux, the initial amount of VLB in CEM and KB-3-1 cells was usually 10- to 20-fold that of CEM/VLB$_{100}$ and KB-VI. In order to compare the rate of efflux from sensitive and resistant cells containing the same, initial amount of VLB, CEM and KB-3-1 cells were loaded with [^{3}H]VLB for only brief periods of time, so that the final loaded concentration was small and comparable to that of MDR cells .

As seen in Fig. 8, at a low cellular concentration there was no

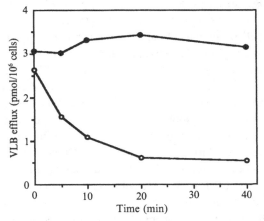

Fig. 8. Efflux of [³H]VLB from drug-sensitive and drug-resistant cells. CEM cells were loaded with [³H]VLB for 30 s so that the level of [³H]VLB in the cells was approximately the same as in their drug-resistant counterparts. CEM/VLB$_{100}$ cells, which were loaded for 30 min (1.07 μM, 0.5 μCi/ml [³H]VLB), were washed and resuspended in TMSucrose containing 5 mM glucose (5 × 10⁶ cells/ml) to study efflux, as described by Warren *et al.* (1991). Radioactivity in 5 × 10⁵ cells in 100 μl aliquots was determined in duplicate samples before and at various times during incubation. Drug-sensitive CEM (●) and drug-resistant CEM/VLB$_{100}$ (○). SEM, <7.2% (*n* = 4).

apparent efflux of [³H]VLB from drug-sensitive cells, while rapid efflux took place from the resistant cells. Adsorbed VLB cannot account for the different efflux curves. The same results were obtained with unwashed VLB-loaded cells as with washed cells. Approximately 5–15% of the initial radioactive VLB was removed by washing. The drug-sensitive cells do not appear to have an effective method of eliminating [³H]VLB below the plateau level, attained after 10–20 min of efflux (Fig. 7). This may be a critical factor in the susceptibility of the cell to drugs.

It may be argued that a rapid efflux of [³H]VLB from

CEM/VLB$_{100}$ cells, presumably effected by P-170, results in a deceptively low rate of influx of drugs into the resistant cell. Drugs can be expelled immediately as they enter the lipid membrane or the cytoplasmic phase. However, if the rate of influx of drugs into the resistant cell is, in fact, the same as into the CEM cell, i.e. 3.2 pmol/10^6 cells per min in our experiments, the rate of efflux from CEM/VLB$_{100}$ cells would have to be far greater than the rate actually observed (~0.20 pmol/10^6 cells per min) to account for an overwhelming of influx. Careful examination of efflux rates shows that, at most, the relative loss of drugs from CEM/VLB$_{100}$ cells is less than 2-fold greater than from CEM cells, even in the 1st min of efflux. [^{3}H]VLB lost from the cell does not re-enter the cell. If during influx of drugs, efflux is taken into consideration, even with very small intracellular concentrations of drugs, accumulation of drugs in CEM cells after only 20–30 s is at least 5–10 times greater than into CEM/VLB$_{100}$ cells. During the first 10 min of efflux, drug-sensitive cells loaded with VLB and containing essentially no P-170 lose seven times more VLB compared to CEM/VLB$_{100}$ cells, under our experimental conditions (Fig. 7). Adsorption of [^{3}H]VLB to both drug-sensitive and drug-resistant cells at 0 °C is relatively small, does not increase with time, and is the same in both cells. This possible source of error in our considerations of influx and efflux of drug is relatively small.

Modulation of multidrug resistance

CEM/VLB$_{100}$ cells, exposed to sodium orthovanadate, verapamil or nigericin, increase uptake of [^{3}H]VLB and a combination of vanadate plus verapamil has an additive effect on the influx of VLB that approaches that of CEM cells (Table 8). These agents have little or no effect on the drug-sensitive CEM cell. Cytotoxicity assays carried out in parallel with the above uptake experiments indicate that vanadate increases cytotoxicity in CEM/VLB$_{100}$ cells to a greater extent than does verapamil, but the two combined do not enhance cytotoxicity additively, suggesting

Table 8. *Effect of drugs on accumulation of vinblastine*

	Change in uptake of [³H]VLB (%)			
Drug	CEM	CEM/VLB$_{100}$	KB-3-1	KB-VI
Verapamil,				
50 μM	−8 ± 13 (5)	+349 ± 47 (6)	−1 ± 98 (7)	+390 ± 51 (7)
Nigericin,				
10 μM	+99 ± 9 (4)	+521 ± 126 (4)	+28 ± 11 (4)	+351 ± 37 (4)
Na vanadate,				
1 μM	+135 ± 12 (7)	+431 ± 96 (7)		
Na vanadate,				
1mM + verapamil	+79 ± 8 (3)	+824 ± 92 (3)		

Notes: Incubation of 4×10^5 cells with [³H]VLB (0.05 μCi) in 100 μl of Tris (10 mM), MgCl$_2$ (1 mM), sucrose (250 mM), pH 7.1 for 10 min at 37 °C. Number of determinations in parentheses.

that cytotoxicity depends not only on the intracellular level of drug but perhaps on its distribution within the cell as well.

Verapamil and other Ca^{2+} channel blockers are known to combine with, and interfere with, the pumping action of P-170 and have been used clinically for reversal of MDR (Tsuruo *et al.*, 1982; Beck, 1987). It is also believed that they interact with membrane lipids (Klohs *et al.*, 1986). The mode of action of vanadate is complex, inhibiting several phosphatases. Vanadate is a well-known inhibitor of phosphorylated ATP pumps and could inhibit P-170 action as well as the passage of drug into isolated, extracellular vesicles (Horio *et al.*, 1988). Whatever the activities of verapamil and vanadate, they appear to be different and their effects on the accumulation of [³H]VLB by MDR cells are additive rather than synergistic (Table 8).

While vanadate enhances accumulation of VLB in CEM/VLB$_{100}$ cells, it does not appear to decrease efflux (Fig. 9) which suggests that it acts by enhancing influx. A similar result is found with nigericin. On the other hand, verapamil does inhibit efflux (Fig. 9). It would appear from these data that the rate of influx of drug is important, although not the sole determinant of accumulation in the MDR cell.

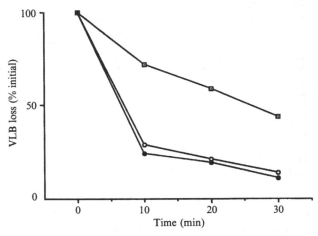

Fig. 9.   Efflux of [³H]VLB from multidrug-resistant CEM/VLB$_{100}$ cells. The same methods were used as in the experiments described in Fig. 7. Controls without modulator (○), in the presence of 1 mM sodium vanadate (●) or 50 μM verapamil (□).

The role of lipids in the accumulation of drugs by MDR cells

As previously discussed, drugs such as vinblastine and the tetra-cyclines appear to enter animal cells by diffusion. It is not unreasonable to believe that during passage of lipophilic drugs through the plasma membrane, important interactions between drug and membrane lipids take place, and that the permeability of the membrane could depend on the packing and the mobility of the membrane lipids. Ling suggested, in 1975, that membrane glycoproteins may modulate membrane fluidity, thus controlling entry of drugs, but direct measurements of membrane fluidity in drug-sensitive and drug-resistant cells have led to contradictory results. However, we have shown that exposure of human leukemic MDR cells CEM/VLB$_{100}$ and K562/ADR to liposomes of specific

Table 9. *Effect of liposomes on uptake of [^{3}H]vinblastine*

Composition of liposome	pmol/10^6 cells[a]			
	CEM		CEM/VLB$_{100}$	
Control	20.0 ± 3.9	(42)	1.6 ± 0.4	(54)
Ph choline (bovine brain)	17.8 ± 5.5	(3)	2.1 ± 0.2	(3)
Ph choline (dipalmityl)	19.1	(1)	2.5	(1)
Ph ethanolamine (bovine brain)	15.1 ± 3.8	(4)	1.7 ± 0.8	(4)
Ph ethanolamine (dioleoyl)	$25.6 + 0.5$	(2)	2.1 ± 0.4	(3)
Ph ethanolamine (distearoyl)	20.2	(1)	0.9	(1)
Ph ethanolamine (dipalmityl)	24.4	(1)	3.1	(1)
Cholesterol	13.7 ± 4.6	(8)	1.6 ± 0.4	(8)
Ph inositol (bovine liver)	19.8 ± 4.8	(9)	16.0 ± 3.1	(4)
Ph inositol + cholesterol (20 mg/ml)	–		7.1	(1)
Ph inositol + cholesterol (50 mg/ml)	9.2	(1)	3.6	(1)
Ph serine (bovine brain)	25.1 ± 0.4	(2)	3.6 ± 0.5	(3)
Phosphatidic acid (dioleoyl)	17.4 ± 1.9	(17)	11.4 ± 2.2	(9)
Phosphatidic acid (arach-stearoyl)	24.9 ± 2.9	(1)	4.4 ± 0.5	(3)
Phosphatidic acid (dimyristyl)	20.3	(1)	4.3	(1)
1,3 diolein	21.7 ± 3.6	(2)	7.9	(1)
Cardiolipin (beef heart)	21.2 ± 4.6	(3)	14.9 ± 0.5	(8)
Cardiolipin (*E. coli*)	24.2 ± 1.0	(3)	2.2 ± 0.5	(5)
Retinol	20.9 ± 2.3	(11)	4.7 ± 1.0	(10)
Retinoic acid	–		1.3	(1)
Retinyl acetate	21.1	(1)	2.6	(1)
Retinyl palmitate	24.6	(1)	1.4	(1)

Notes: [a]Mean ± SEM. Numbers in parentheses are the number of experiments, each assay in duplicate.
The liposomes used were made by a standard method. Lipids, in chloroform-methanol (9:1) or dimethylsulfoxide, were dried under a stream of N_2 in a glass tube. TMSucrose buffer was added to the tube, which was ultrasonicated for 30 s at room temperature in a Branson ultrasonicator cleaner to form large multilamellar vesicles. The preparation retained its activity after filtration through polycarbonate filters with 0.1 or 0.2 micropores. Liposomes could be stored for at least 1 week at 4 °C without loss of activity. Influx assay (10 min) as described in Warren *et al.* (1991). Ph, phosphatidyl.

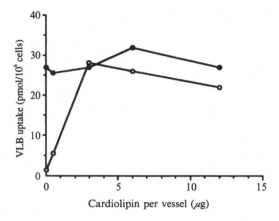

Fig. 10. Effect of cardiolipin liposomes (beef heart) on the accumulation of [³H]VLB by CEM (drug-sensitive)(●) and CEM/VLB$_{100}$ (drug-resistant) (○) cells. Procedures described in Warren *et al.* (1992). One of three experiments with similar results.

composition significantly increases the accumulation and cytotoxicity of [³H]VLB, while there is relatively little effect on accumulation of drug by drug-sensitive cells. We believe that treatment of cells with liposomes before adding drug results in a fusion of cell membrane and liposome, changing the composition of the plasma membrane. A change in some unknown property of the surface membrane, in which P-170 is embedded, renders the membrane penetrable by drugs (Warren *et al.*, 1992).

Accumulation of [³H]VLB can be raised in CEM/VLB$_{100}$ cells exposed to liposomes made from cardiolipin, phosphatidyl inositol, and dioleoyl phosphatidic acid by a standard method (Table 9).

Although unsaturated fatty acids are necessary, they are not sufficient, since phosphatidylcholine- and phosphatidylethanolamine-bearing unsaturated fatty acids are inactive. Beef heart cardiolipin with over 90% unsaturated fatty acids (> 80% C-18:2) is stimulatory while bacterial cardiolipin (*E. coli*) with fatty acids

Table 10. *Effect of liposomes on cytotoxicity of vinblastine*

Liposomes or addition	IC$_{50}$ (M)^a	n
None	$3.15 \pm 0.45 \times 10^{-4}$	4
Cardiolipin	$2.35 \pm 0.43 \times 10^{-5}$	4
Phosphatidylinositol	$1.35 \pm 0.07 \times 10^{-4}$	3
Dioleoyl phosphatidic acid	$1.20 \pm 0.17 \times 10^{-4}$	3
Verapamil (50 mM)	$1.65 \pm 0.26 \times 10^{-4}$	3

Notes: Each concentration of VLB was tested in triplicate. IC$_{50}$, the concentration of VLB that inhibits cell growth by 50% when compared to untreated controls. Experimental data were analyzed according to the Student's *t* test.
^{a}Results are given as mean ± SEM where n is the number of experiments.

that are mostly saturated (45% C-16, 20% C-18), is inactive. A negative charge contributes to activity, although 1,3-diolein is partially effective; on the other hand, phosphatidylserine is almost inactive. Cholesterol, either alone or with phosphatidylinositol, inhibits uptake by drug-sensitive cells and can abolish the stimulatory effect of phosphatidylinositol in CEM/VLB$_{100}$ cells.

Figure 10 shows that the liposome effect increases with lipid concentration (cardiolipin) to a point where accumulation of [^{3}H]VLB is approximately the same in CEM and CEM/VLB$_{100}$ cells. There is virtually no effect on drug accumulation by CEM cells.

As expected, liposomes increase the cytotoxicity of VLB in CEM/VLB$_{100}$ cells (Table 10). Cardiolipin liposomes at a concentration of 100 μg/ml increase the toxicity over 10 times, which is far more effective than verapamil. Liposomes made of phosphatidylinositol and dioleoyl phosphatidic acid are not as active as cardiolipin liposomes.

The liposomes themselves appear to be bound to the cells. CEM/VLB$_{100}$ cells have been incubated with cardiolipin or phosphatidylinositol liposomes for 10 min, and then centrifuged at low *g* force to separate the remaining, free liposomes from cells. The supernatant is removed. When cells are resuspended in fresh buffer and assayed, stimulation of uptake of [³H]VLB by liposomes, which must be associated with the cells, is the same as when the original liposome preparation is present. Maximum stimulation occurs after 10 min of preincubation of liposomes and cells. The rapidity of the process suggests that lipid exchange does not underlie stimulation. Resuspended cells, previously exposed to cardiolipin liposomes and incubated at 37 °C for different periods of time have been assayed for [³H]VLB uptake to see how long the liposome stimulatory effect lasts. The effect declines with a half-life of 1.5–2.2 h (n=3).

Using phosphatidylinositol liposomes containing [³H]phosphatidylinositol, it has been found in a typical assay that 10^6 CEM and CEM/VLB$_{100}$ cells pick up 0.31–0.58 μg lipid (n = 3). Uptake is approximately the same in drug-sensitive and MDR cells, both of which contain 5 μg phospholipid per 10^6 cells.

Our experiments have shown that liposomes in this system did not pick up [³H]VLB and deliver it to the cell. To demonstrate this, liposomes were incubated with the same amount of [³H]VLB as used in a typical assay, in the absence of cells. After 10 min, the liposomes were centrifuged (15 000 g, 10 min, 4 °C) and the supernatant containing soluble [³H]VLB was removed. The pellet was resuspended and exposed to cells in the standard assay. Uptake of [³H]VLB by cells was less than 16% of that in the usual assay (incubation of cells, liposomes and [³H]VLB). When the liposome pellet was washed once with TMSucrose buffer at 4 °C, the value was reduced to 7%.

Exposure of CEM/VLB$_{100}$ cells to anti P-170 monoclonal antibody MRK-16 (1 μg/ml) that is specific for an epitope on the cell surface (Hamada and Tsuruo, 1986), before incubation with liposomes, does not alter the stimulatory response to liposomes in [³H]VLB uptake assays. This suggests that liposomes do not combine directly with P-170, and reduces the possibility that liposomes are acting as an agonist that combines with P-170 functioning as a receptor. One interpretation of the data is that the liposomes rapidly fuse with the plasma membrane of the cells and change its composition. Fusion of liposomes with cells is known to take place and has been studied extensively (Pagano and

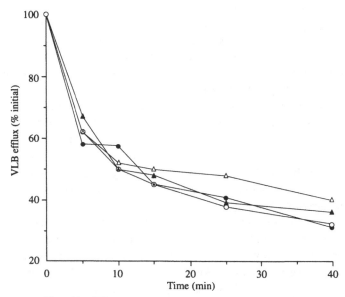

Fig. 11. Effect of liposomes on efflux of [³H]VLB from CEM/VLB₁₀₀ cells. The procedure is described in Warren *et al.* (1992). Initial 100% values (time, 0) expressed as pmol/10⁶ cells are as follows: control, 1.4 ± 0.3; cardiolipin, 13.8 ± 4.0; dioleoyl phosphatidic acid, 11.9 ± 2.8; phosphatidylinositol, 15.8 + 2.3 (*n* = 3). Control (O); beef heart cardiolipin, 100 µg/ml (●); dioleoyl phosphatidic acid, 200 µg/ml (△); phosphatidylinositol, 200 µg/ml (▲).

Weinstein, 1978). If this is so, change in composition of the lipids of the plasma membrane might inhibit the efflux pump, as do verapamil and other Ca²⁺ channel blockers, calmodulin inhibitors and Tween-80 (Klohs *et al.*, 1986). However, as seen in Fig. 11, efflux of [³H]VLB from liposome-treated CEM/VLB₁₀₀ cells is not significantly changed, which has prompted us to direct our attention to the possible influence of the modified plasma membrane on influx of [³H]VLB. The rate of efflux is not altered by liposomes even when measured after only 1 min.

Is it possible that P-170, each molecule of which passes through the plasma membrane 12 times, interacts extensively with lipid, changing the conformation, packing and/or motion of lipid molecules so that traversal by lipophilic drugs is reduced? If so, our results indicate that by changing the composition of the surface membrane of the MDR cell, the permeability to drugs reverts to that of the drug-sensitive cell.

McElhaney, de Gier and van der Neut-Kok, (1973) and de Kruyff *et al.* (1973) have shown that the passage by diffusion of glycerol and erythritol into *Acholeplasma laidlawii* cells could be altered by changing the fatty acids and the cholesterol content in its membrane by diet. Reduced length, branching and unsaturation of the fatty acids increase non-electrolyte permeability. Cholesterol decreases permeability. Our results appear to be similar to these. The effects can be explained by changes in fluidity of membrane lipids. Since diffusion through a membrane depends on the packing density and motion of the lipids, increased order might reduce entry of drugs into a drug-resistant cell. An increased order might be induced by extensive P-170–lipid interaction and an increase in fluidity in our cell system could be brought about by introducing into the membrane certain specific phospholipids containing unsaturated fatty acids which, we suggest, would restore considerable permeability to drugs. The introduction of cholesterol into the membrane, which is known to decrease fluidity, inhibits accumulation of [^{3}H]VLB by both drug-sensitive and phosphatidylinositol liposome-stimulated resistant cells (Table 9). However, studies on the fluidity of membrane lipids of drug-sensitive and multidrug-resistant cells have yielded conflicting data, and to date, systematic shifts in lipid order in MDR cells have not been consistently observed.

We believe that, in this study, liposomes do not entrap a drug in order to 'deliver' the drugs to a cell, an approach that has been extensively investigated for improving therapy in general and also for enhancing introduction of anti-cancer drugs into MDR cells (Thierry *et al.*, 1989; Fan *et al.*, 1990; Sadasivan *et al.*, 1991). We believe they operate by changing the composition (and permeability) of the plasma membrane.

Whatever the cellular mechanism of stimulation, liposomes may prove useful in the process of overcoming multidrug-resistance. As shown in Table 10, liposomes of cardiolipin, phosphatidylinositol and dioleoyl phosphatidic acid clearly increase the toxicity of VLB in CEM/VLB$_{100}$ cells, even more than verapamil. The effective lipids are nutrients and, being normal body constituents, should not be particularly toxic, as are so many pharmacological agents that are being used for reversing multidrug-resistance. Lipids of other fatty acid composition, or different combinations of lipids, may prove more efficacious for specific cells.

P-170 and membrane lipids

Evidence has been obtained supporting the idea that the glycoprotein P-170 in the surface membrane, an efflux pump, is also capable of reducing passage of lipophilic compounds. How can this be brought about? As previously stated, each molecule of P-170 possesses 12 transmembrane segments, so that interaction between the lipid bilayer and the hydrophobic amino acids must be extensive. The possibility exists that the α-helical transmembrane sequences induce an order or particular arrangement of membrane lipids that is unfavorable for the passage of lipophilic molecules through the lipid bilayer. Although rough estimates of what fraction of the cell surface may be covered with P-170 are as high as 10% (Riordan and Ling, 1985), it could be considerably less, so that there simply may only be enough transmembrane segments of P-170 per unit area to affect a small fraction of the total lipid bilayer. Furthermore, each transmembrane segment may influence only a few rows of surrounding lipid molecules, so that the total, altered surface area of the cell through which lipophilic drugs diffuse may not be sufficient to account for the observed, reduced influx in MDR cells.

Another possibility is that a significant fraction of drug entering the lipid bilayer is attracted to the transmembrane segment. While some drug passes through the membrane, much of it, interacting

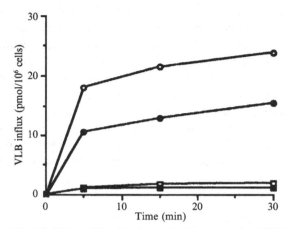

Fig. 12. Effect of Sendai virus infection on uptake of [³H]VLB. CEM and CEM/VLB$_{100}$ cells (5×10^6) in loose pellets were exposed to 2×10^7 infectious Sendai virus particles for 30 min at 37 °C. After diluting and washing, the cells were cultured for 15 h, at which time there is full expression of Sendai virus glycoprotein on the cell surface. Influx of [³H]VLB in cells was assayed in duplicate ($n = 3$). CEM, ○; CEM-virus, ●; CEM/VLB$_{100}$, □; CEM/VLB$_{100}$-virus, ■. Virus kindly provided by H. Ertl.

with hydrophobic amino acids, is conducted back to the cell surface along the hydrophobic polypeptide, perhaps by a flipping mechanism (Higgins and Gottesman, 1992); this is not necessarily an energy-requiring, pumping activity. Therefore, P-170 may function both as an energy-requiring, efflux pump and as a short-circuiting conduit for drugs by virtue of its hydrophobic trans-membrane sequences. This notion was formulated when we found that the rate of entry of [³H]VLB and [³H]daunomycin (DM) into several cells bearing relatively large amounts of specific surface membrane glycoproteins with transmembrane domains was significantly less than into cells with little or none of these glyco-proteins. We have studied several systems:

(a)   CEM cells before and after infection with Sendai virus in

Table 11. *Cytotoxicity of vinblastine and daunomycin in human liver cells*

	IC_{50} (M)	
	PLC	NPLC
Vinblastine	$3.1 (\pm 0.2) \times 10^{-6}$	$1.7 (\pm 0.1) \times 10^{-5}$
Daunomycin	$1.2 (\pm 0.2) \times 10^{-5}$	$1.0 (\pm 0.2) \times 10^{-4}$

Notes: IC_{50} is the concentration of drugs that inhibits growth by 50% after 48 h. Each concentration was tested in duplicate in 24-well plates. Cells were counted with a Coulter counter. Mean $\pm$ SEM ($n = 3$).

which, after 15 h, a viral glycoprotein with one transmembrane segment per molecule coats the infected cell surface (Fig. 12).

(b) A Chinese hamster ovary cell before and after transfection with a gene coding for the β-adrenergic receptor (O'Dowd, Lefkowitz and Caron, 1989) (Fig. 13).

(c) An NIH3T3 cell before and after transfection with a gene coding for a serotonin receptor (10^5–10^6 receptors per cell), seven transmembrane segments per molecule (Julius *et al.*, 1988) (Fig. 14).

(d) A human liver cell PLC with 5×10^5 EGF receptors per cell and a second line (NPLC) with 2×10^6 EGF receptors per cell (Carlin *et al.*, 1988) (Fig. 15).

The rate of entry of VLB or daunomycin into the cells with increased transmembrane sequences was clearly reduced and yet none of these surface glycoproteins has ever been known to function as a pump. Not only is the accumulation of drugs reduced in NPLC, as expected, but also the cytotoxicity of the drugs is decreased compared to that of PLC (Table 11). Myers *et al.* (1986) have shown that multidrug-resistant Chinese hamster and mouse tumor cells with overexpressed P-170 bear increased

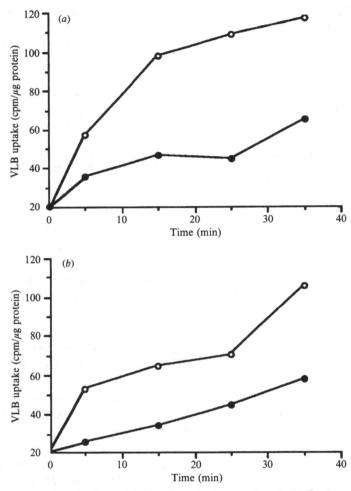

Fig. 13. Control CHO cells and cells transfected with β-adrenergic receptor gene, grown in 30 mm, 6-well plastic dishes,
were washed with cold TMSucrose, and then incubated for
different periods of time in TMSucrose containing (*a*) 0.05 μCi
[³H]vinblastine (1.07 nmoles); or (*b*) 0.05 μCi [³H]daunomycin
(1 nmole). Cells were washed once with cold TMSucrose,
scraped, and transferred to vials with scintillation fluid (5 ml)

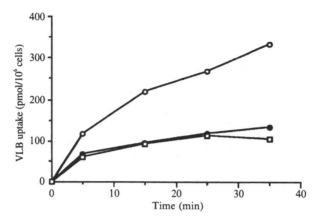

Fig. 14. Uptake of [³H]VLB by NIH3T3 cells and two lines of these cells transfected with the serotonin receptor gene (8×10^5 receptors per cell, seven transmembrane segments per molecule). Uptake of drug measured by the procedure described in Fig. 13. NIH3T3, ○; 5HTIC, □; 5HT2, ●. We thank Dr David Julius for the cells (Julius *et al.*, 1988).

amounts of epidermal growth factor receptors. Our results indicate that the receptors, without P-170, also confer resistance. PLC and NPLC cells bind very small and equal amounts of [¹²⁵I] MRK16, an antibody directed against an epitope of P-170 on the cell surface (Hamada and Tsuruo, 1986). The amount of ¹²⁵I-labeled antibody bound was approximately the same as that of drug-sensitive NIH3T3 cells, which we consider a background value.

These results suggest that an unexpected consequence of an

Caption for Fig. 13 (*cont*).

and the radioactivity determined. Control cells, ○; transfected cells, ●, expressing β-adrenergic receptors. The results are expressed as uptake of drug per μg cell protein. The same result was obtained calculating results on a per cell basis. Each point is the mean of triplicate determinations ($n = 3$). We thank Dr R.J. Lefkowitz for supplying the cells (O'Dowd *et al.*, 1989).

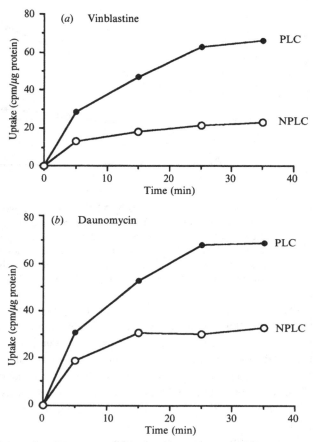

Fig. 15. Uptake of [³H]VLB and [³H]daunomycin by human liver cells with low (PLC) and high (NPLC) levels of epidermal growth factor receptors in their surface membranes (5×10^5 *vs* 2×10^6 receptors per cell). The procedure was the same as described in the legend of Fig. 13. PLC, ●; NPLC, ○. We thank Dr Barbara B. Knowles for the cells (Carlin *et al.*, 1988).

enrichment of transmembrane glycoproteins (viruses, receptors, etc.) may be an altered permeability of the cell to lipophilic and perhaps other unknown substances.

The association of reduced drug uptake with increased transmembrane peptides, found in the cells listed above, does not always hold. For instance, uptake of [^{3}H]VLB and [^{3}H]daunomycin is precisely the same in human embryonic kidney cells before and after transfection with the bovine rhodopsin gene (Nathans *et al.*, 1988). A transfected cell bearing 2×10^6 rhodopsin molecules takes up drugs as effectively as an untransfected cell. Chinese hamster ovary cells expressing relatively large amounts of rabies virus glycoprotein (Burger *et al.*, 1991) and glycophorin A also take up as much drug as controls. It should be noted that the rate of accumulation of drugs by control and transfected cells is precisely the same, in contrast to the first group where there is a significant decrease. In no instance has there been an increase in uptake associated with increased transmembrane glycoproteins. The reason for this lack of effect is unknown, but could lie in the nature of the transmembrane sequences, or in the flanking amino acids, or in the relationships and associations of the transmembrane peptides with one another or with surrounding lipids.

Overview

In conclusion, I have attempted to show that the important phenomenon of multidrug resistance cannot be ascribed solely to the efflux pumping action of P-glycoprotein, although it is a very important component. Some workers believe that a differential binding of drugs to cytoplasmic structures may be important and others place great emphasis on the different distribution of drugs within drug-sensitive and drug-resistant cells.

In this summary of our data, we have argued that decreased influx of drugs induced by P-170 in MDR cells may play a critical role and that a balance of decreased influx and increased efflux, depending on the type of cell and its condition and on the structure of the drug, may describe drug-resistance more accurately. Our emphasis on decreased influx is based on a number of observations:

(a)   In several MDR cells, the kinetic data suggest that while the rate of efflux is greater in the MDR cell it is not sufficient to account for the observed 10- to 20-fold decreases in accumulation of drugs by MDR cells, even after only 20 s of influx. Unfortunately, kinetic experiments are difficult, and it is possible that drugs are expelled by P-170 immediately upon entering the surface membrane. However, we have found that when cells are loaded with VLB or daunomycin and the entry of other ^{3}H-labeled drugs are then measured, entry is unaffected by the presence of a 10-fold greater concentration of intracellular drugs. Surely, the intracellular drug would have a selective, competitive advantage for the efflux pump over entering material. If the intracellular material occupied the efflux pump, the apparent accumulation of ^{3}H-labeled drug (VLB or daunomycin) would increase: this is not observed. However, many uncertainties remain in attempting to explain with kinetic data a decreased accumulation of drugs by MDR cells because rates of influx and efflux must be considered together with the variable binding of drugs both outside and inside the lipid bilayer of the surface membrane.

(b)   Our work with liposomes suggests, but does not prove, that altering the lipid composition of the surface membrane of MDR cells changes some physical property induced by P-170 that is responsible for decreased entry (diffusion) of drugs into the cell. Efflux of drugs, the known function of P-170, is apparently unaltered.

(c)   Numerous domains of hydrophobic amino acids spanning the membrane may interfere with passage of drugs, quite independent of a pumping process. This may be so with P-170 although several transmembrane glycoproteins do not

decrease the rate of accumulation of drugs. Why some do
and others do not (but none stimulates drug accumulation)
remains to be answered. Perhaps only the active membrane
glycoproteins affect lipid order or fluidity.

References

Ada, G.L. and Lind, P.E. (1961). *Nature* **190**, 1171.

Aminoff, D. (1961). *Biochem. J.* **81**, 384.

Arango, J., Shoreibak, M. and Pierce, M. (1987). In *Proc. IX Int. Symp. Glycoconjugates*, Abstr. E114.

Ashwell, G. and Morell, A.G. (1981). *Ann. Rev. Biochem.* **51**, 531.

Atterfelt, P., Blohme, I., Norby, A. and Svennerholm, L. (1958). *Acta Chem. Scand.* **12**, 359.

Baker, S.R., Blithe, D.L., Buck, C.A. and Warren, L. (1980). *J. Biol. Chem.* **255**, 8719.

Beau, J.-M. and Schauer, R. (1980). *Eur. J. Biochem.* **106**, 531.

Beaudet, A.L. (1983). In *The Metabolic Basis of Inherited Diseases*, 5th edn, eds. J.B. Stanbury, J.B. Wyngaarden, D.S., Fredrickson, J.L. Goldstein and M.S. Brown, p. 788. New York: McGraw-Hill.

Beck, W.T. (1987). *Biochem. Pharmacol.* **36**, 2879.

Beintema, J.J., Gastra, W., Scheffer, A.J. and Wetting, G.W. (1976). *Eur. J. Biochem.* **63**, 441.

Benkovic, S.J., Adams, J.A., Borders, C.L. Jr, Janda, K.D. and Lerner, R.A. (1990). *Science* **250**, 1132.

Bernacki, R.J., Niedbala, M.J. and Korytnyk, W. (1985). *Cancer Metastasis Rev.* **4**, 81.

Blacklow, R.S. and Warren, L. (1962). *J. Biol. Chem.* **237**, 3520.

Blix, G. (1936). *Z. Physiol. Chem.* **240**, 43.

Bose, S., Gurari-Rotman, D., Ruegg, U.T., Corley, L. and Anfinsen, C.B. (1976). *J. Biol. Chem.* **251**, 1659.

Bosmann, H.B. (1972). *Biochem. Biophys. Res. Commun.* **49**, 1256.

Bosmann, H.B. and Kessel, D. (1970). *Mol. Pharmacol.* **6**, 345.

Buck, C.A., Glick, M.C. and Warren, L. (1971). *Biochem.* **10**, 2176.

Burger, S.R., Remaley, A.T., Danley, J.M., Moore, J., Muschel, R.J., Wunner, W.H. and Spitalnik, S.L. (1991). *Gen. Virol.* **72**, 359.

Burnet, F.M. and Stone, J.D. (1947). *Aust. J. Exp. Biol. Med. Sci.* **25**, 227.

Burt, R.K., Fojo, A.T. and Thorgeirsson, S.S. (1990). *Hospital Practice* **25**, 67.

Carlin, C.R., Simon, D., Mattison, J. and Knowles, B.B. (1988). *Mol. Cell Biol.* **8**, 25.

Cheng, S., Morrone, S. and Robbins, J. (1979). *J. Biol. Chem.* **254**, 8830.

Choi, K., Chen, C., Kriegler, M. and Roninson, I.B. (1988). *Cell* **53**, 519.

Clamp, J. (1974). *Biochem. Soc. Symp.* **40**, 3.

Comb, D.G. and Roseman, S. (1958a). *J. Biol. Chem.* **232**, 807.

Comb, D.G. and Roseman, S. (1958b). *J. Am. Chem. Soc.* **80**, 497.

Comb, D.G. and Roseman, S. (1960). *J. Biol. Chem.* **235**, 2529.

Corfield, A.P. and Schauer, R. (1982). In *Sialic Acids, Chemistry Metabolism and Function*, p. 6. Wien: Springer Verlag.

Cossu, G. and Warren, L. (1983). *J. Biol. Chem.* **258**, 5603.

Cossu, G., Kuo, A., Pessano, S., Warren, L. and Cooper, R.A. (1982a). *Cancer Res.* **42**, 484.

Cossu, G., Warren, L., Boettiger, D., Holtzer, H. and Pacifici, M. (1982b). *J. Biol. Chem.* **257**, 4463.

Currier, S.J, Veda, K., Willingham, M.C., Pastan, I. and Gottesman, M.M. (1989). *J. Biol. Chem.* **264**, 14376.

de Kruyff, B., de Greef, W.J., van Eyck, R.V.W., Demel, R.A. and van Deenen, L.L.M. (1973). *Biochim. Biophys. Acta* **298**, 479.

Dennis, J.W. (1988). *Cancer Surveys* **7**, 573.

Dennis, J.W. and Laferte, S. (1989). *Cancer Res.* **49**, 945.

Dennis, J.W., Laferte, S., Yagel, S. and Breitman, M.L. (1989). *Cancer Cells* **1**, 87.

Dietzschold, B. (1977). *J. Virol.* **23**, 286.

Endicott, J.A. and Ling, V. (1989). *Ann. Rev. Biochem.* **58**, 137.

Etchison, J.R. and Holland, J.J. (1974). *Proc. Natl. Acad. Sci. USA* **71**, 4011.

Eylar, E.H. (1965). *J. Theor. Biol.* **10**, 89.

Fan, D., Bucana, C.D., O'Brian, C.A., Zwelling, L., Seid, C. and Fidler, I.J. (1990). *Cancer Res.* **50**, 3619.

Fenderson, B.A., Ostrander, G.K., Hausken, Z., Radin, N.S. and Hakomori, S.I. (1992). *Exp. Cell Res.* **198**, 362.

Finne, J. (1985). *TIBS* **10**, 129.

Finne, J. and Mäkela, P.H. (1985). *J. Biol. Chem.* **260**, 1265.

Fojo, A.T., Shen, D.-W., Mickley, L.A., Pastan, I. and Gottesman, M.M. (1987). *J. Clin. Oncol.* **5**, 1922.

Fontaine, G., Biserte, G., Montreuil, J., DuPont, A. and Farriaux, J.P. (1968). *Helv. Paedeatr. Acta* **23**, (suppl. XVII), 3.

Foote, S.J., Thompson, J.K., Cowman, A.F. and Kemp, D.J. (1989). *Cell* **57**, 921.

Ford, J.M. and Hait, W.N. (1990). *Pharmacol. Rev.* **42**, 155.

Ghosh, S. and Roseman, S. (1965a). *J. Biol. Chem.* **240**, 1531.

Ghosh, S. and Roseman, S. (1965b). *J. Biol. Chem.* **240**, 1525.

Glick, M.C. and Buck, C.A. (1973). *Biochem.* **12**, 85.

Glick, M.C., Rabinowitz, Z. and Sachs, L. (1973). *Biochem.* **12**, 4864.

Glick, M.C., Rabinowitz, Z. and Sachs, L. (1974). *J. Virol.* **13**, 967.

Gottesman, M.M. and Pastan, I. (1988). *J. Biol. Chem.* **263**, 12163.

Gottschalk, A. (1960). *The Chemistry and Biology of Sialic Acids and Related Substances.* London: Cambridge University Press.

Gottschalk, A. (1972). In *Glycoproteins*, vol. 5A, ed. A. Gottschalk, pp. 1–23. Amsterdam: Elsevier.

Gottschalk, A. and Lind, P.E. (1949). *Nature (Lond.)* **164**, 232.

Gottschalk, A. and Thomas, M.A.W. (1961). *Biochim. Biophys. Acta.* **46**, 91.

Hakomori, S. and Kannagi, R. (1983). *J. Natl. Canc. Inst.* **71**, 231.

Hamada, H. and Tsuruo, T. (1986). *Proc. Natl. Acad. Sci. USA* **83**, 7785.

Hart, G.W., Haltiwanger, R.S., Holt, G.D. and Kelly, W.G. (1989). *Ann. Rev. Biochem.* **58**, 841.

Heimer, R. and Meyer, K. (1956). *Proc. Natl. Acad. Sci. USA* **42**, 728.

Hentschel, H. and Muller, M. (1979). *Comp. Biochem. Physiol.* **64A**, 585.

Herrier, G., Rott, R., Klenk, H.-D., Muller, H.-P., Shukla, A.K. and Schauer, R. (1985). *EMBO J.* **4**, 1503.

Higa, H.H., Butor, C., Diaz, S. and Varki, A. (1989). *J. Biol. Chem.* **264**, 19427.

Higgins, C.F. and Gottesman, M.M. (1992). *TIBS* **17**, 18.

Hirschberg, C.B. (1987). *Ann. Rev. Biochem.* **56**, 63.

Hirschberg, C.B., Wolf, B.H. and Robbins, P.W. (1974). *J. Cell Physiol.* **85**, 31.

Horio, M., Gottesman, M.M. and Pastan, I. (1988). *Proc. Natl. Acad. Sci. USA* **85**, 3580.

Hunt, L.A. and Summers, D.F. (1976a). *J. Virol.* **20**, 637

Hunt, L.A. and Summers, D.F. (1976b). *J. Virol.* **20**, 647.

Irani, R.J. and Ganapathi, K. (1962). *Nature (Lond.)* **195**, 1227.

Jakoby, R.K. and Warren, L. (1961). *Neurology* **11**, 232.

Jaques, L.W., Brown, I.B., Barrett, J.M., Brey, W.S. Jr and Weltner, W. Jr (1977). *J. Biol. Chem.* **252**, 4533.

Jourdian, G.W. and Roseman, S. (1963). *Ann. NY Acad. Sci.* **106**, 202.

Julius, D., MacDermott, A.B., Jassell, T.J. and Axel, R. (1988). *Science* **241**, 558.

Juranka, P.F., Zastawny, R.L. and Ling, V. (1989). *FASEB J.* **3**, 2583.

Kaufmann, S.H.E., Schauer, R. and Hahn, H. (1981). *Immunobiology* **160**, 184.

Kean, E. L. (1991). *Glycobiology* **1**, 441.

Klenk, E. (1941). *Z. Physiol. Chem.* **268**, 50.

Klohs, W.D. and Steinkampf, R.W. (1988). *Mol. Pharmacol.* **34**, 180.

Klohs, W.D., Steinkampf, R.W., Havlick, M.J. and Jackson, R.J. (1986). *Cancer Res.* **46**, 4352.

Kobata, A. (1984). In *Biology of Carbohydrates*, eds. V. Ginsberg and P. W. Robbins, p. 87. New York: John Wiley.

Kobata, A. (1988). *Biochimie* **70**, 1575.

Koprowski, H. (1987). *4th Eur. Symp. Carbohydrates*, Abstr.

Kornfeld, S. (1986). *J. Clin. Invest.* **77**, 1.

Kornfeld, S. (1987). *FASEB J.* **1**, 462.

Kornfeld, R. and Kornfeld, S. (1985). *Ann. Rev. Biol.* **54**, 631.

Kornfeld, S., Kornfeld, R., Neufeld, E. and O'Brien, P.J. (1964). *Proc. Natl. Acad. Sci. USA* **52**, 371.

Krantz, M.J., Lee, Y.C. and Hung, P.P. (1974). *Nature (Lond.)* **248**, 684.

Krogstad, D.J., Gluzman, I.Y., Kyle, D.E. *et al.* (1987). *Science* **238**, 1283.

Lehrman, M.A. (1991). *Glycobiology* **1**, 553.

Ling, V. (1975). *Can. J. Genet. Cytol.* **17**, 503.

Liotta, L. (1986). *Cancer Res.* **46**, 1.

Luppi, A. and Cavazzini, G. (1966). *Nuovi. Ann. Ig Microbiol.* **17**, 183.

Lush, I.E. (1961). Nature **189**, 981.

Lush, I.E. and Conchie, J. (1966). *Biochim. Biophys. Acta* **130**, 81.

Mancini, G.M.S., Verheijen, F.W. and Galjaard, H. (1986). *Hum. Genet.* **73**, 214.

Martin, S.K., Oduola, A.M.J. and Milhous, W.K. (1987). *Science* **235**, 899.

McCrea, J.F. (1947). *Aust. J. Exp. Biol. Med. Sci.* **25**, 127.

McElhaney, R.N., de Gier, J. and van der Neut-Kok, E.C.M. (1973). *Biochim. Biophys. Acta* **298**, 500.

Megaw, J.M. and Johnson, L.D. (1979). *Proc. Soc. Exp. Biol. Med.* **161**, 60.

Miyagi, T. and Tsuiki, S. (1985). *J. Biol. Chem.* **260**, 6710.

Mizuochi, T., Amano, J. and Kobata, A. (1984). *J. Biochem.* **95**, 1209.

Montreuil, J., Biserte, G., Strecker, G., Spik, G., Fontaine, G. and Farriaux, J.-P. (1968). *Clin. Chim. Acta* **21**, 61.

Muramatsu, T., Atkinson, P.H., Nathenson, S.G. and Ceccarini, G. (1973). *J. Mol. Biol.* **80**, 781.

Myers, M.B., Merluzzi, V.J., Spengler, B.A. and Biedler, J.L. (1986). *Proc. Natl. Acad. Sci. USA* **83**, 5521.

Nathans, J., Weitz, C.J., Agarwal, N., Nir, I. and Papermaster, D.S. (1989). *Vision Res.* **29**, 907.

O'Dowd, B.F., Lefkowitz, R.J. and Caron, M.G. (1989). *Ann. Rev. Neurosci.* **12**, 67.

Pagano, R.E. and Weinstein, J.N. (1978). *Ann. Rev. Biophys. Bioeng.* **7**, 435.

Papadopoulos, N.M. and Hess, W.C. (1960). *Arch. Biochem. Biophys.* **88**, 167.

Pattabiraman, T.N. and Bachhawat, B.K. (1961). *Biochim. Biophys. Acta* **54**, 272.

Phillips, M.L., Nudelman, E., Gaeta, F.C.A, Perez, M., Singhal, A.K., Hakomori, S.-I. and Paulson, J.C. (1990). *Science* **250**, 1130.

Plummer, T.H. Jr and Hirs, C.H.W. (1964). *J. Biol. Chem.* **239**, 2530.

Rahmann, H. and Breer, H. (1976). *Roux's Arch. Develop. Biol.* **180**, 253.

Renlund, M., Tietze, F. and Gahl, W.A. (1986). *Science* **232**, 759.

Riordan, J.R. and Ling, V. (1985). *Pharmacol. Ther.* **28**, 51.

Riordan, J.R., Rommens, J.M., Kerem, B. *et al.* (1989). *Science* **245**, 1066.

Roseman, S. (1962). *Proc. Natl. Acad. Sci. USA* **48**, 437.

Roseman, S., Jourdian, G.W., Watson, D. and Rood, R. (1961). *Proc. Natl. Acad. Sci. USA* **47**, 958.

Roth, J.C., Zuber, C., Wagner, P., Taatjes, D.J., Weisgerber, C., Heitz, P.U., Goridis, C. and Bitter-Suermann, D. (1988). *Proc. Natl. Acad. Sci. USA* **85**, 2999.

Rothman, R.J., Perussia, B., Herlyn, D. and Warren, L. (1989). *Molec. Immunol.* **26**, 1113.

Rutishauser, V., Acheson, A., Hall, A.K., Mann, D.M. and Sunshine, J. (1988). *Science* **24**, 53.

Sadasivan, R., Morgan, R., Fanian, C. and Stephens, R. (1991). *Cancer Lett.* **57**, 165.

Saifer, A. and Siegel, H.A. (1959). *J. Lab. Clin. Med.* **53**, 474.

Schacter, H. (1991). *Glycobiology* **1**, 453.

Schacter, H. and Roseman, S. (1980) In *The Biochemistry of Glycoproteins and Proteoglycans*, ed. W.J. Lennzar, p. 85. New York: Plenum Press.

Schauer, R. (ed.) (1982). *Sialic Acids, Chemistry, Metabolism and Function*, p. 6. Wein: Springer Verlag.

Schauer, R. (1991). *Glycobiology* **1**, 449.

Schengrund, C., Rosenberg, A. and Repman, M.A. (1976). *J. Cell Biol.* **79**, 555.

Schloemer, R.H. and Wagner, R.R. (1974). *J. Virol.* **14**, 270.

Schloemer, R.H. and Wagner, R.R. (1975). *J. Virol.* **15**, 1029.

Seed, T.M., Aikawa, M., Sterling, C. and Rabbege, J. (1974). *Infect. Immun.* **9**, 750.

Sehested, M., Skovsgaard, T., van Deurs, B. and Winter-Nielsen, H. (1987). *J. Natl. Cancer Inst.* **78**, 171.

Seppala, R., Tietze, F., Krasnewich, D. *et al.* (1991). *J. Biol. Chem.* **266**, 7456.

Shames, S.L., Simon, E.S., Christopher, C.W., Schmid, W., Whitsides, G.M. and Yang, L.-L. (1991). *Glycobiology* **1**, 187.

Shaw, L. and Schauer, R. (1988). *Hoppe-Seyler's Z. Physiol. Chem.* **369**, 477.

Sinohara, H. (1972). *Tohoku J. Exp. Med.* **106**, 93.

Smets, L.A. and van Beek, W.P. (1984). *Biochim. Biophys. Acta* **738**, 237.

Strauss, J.H. Jr, Burge, B.W. and Darnell, J.E. (1970). *J. Mol. Biol.* **47**, 437.

Struck, D. K. and Lennarz, W.J. (1980) In *The Biochemistry of Glycoproteins and Proteoglycans*. ed. W. J. Lennarz, p. 35. New York: Plenum Press.

Sullivan, C.W. and Volcani, B.E. (1974). *Arch. Biochem. Biophys.* **163**, 29.

Sylven, B. and Bois-Svensson, I. (1965). *Cancer Res.* **25**, 458.

Tarentino, A.L. and Maley, F. (1974). *J. Biol. Chem.* **249**, 811.

Tarentino, A.L., Gomez, C.M. and Plummer, T.H. Jr (1985). *Biochem.* **24**, 4665.

Thierry, A.R., Jorgansen, T.G., Fozst, D., Delli, C.A., Dritschilo, A. and Rahman, A. (1989). *Cancer Commun.* **1**, 311.

Thurin, J., Herlyn, M., Hindsgaul, O. *et al.* (1985). *J. Biol. Chem.* **260**, 14556.

Tomita, M. and Marchesi, V.T. (1975). *Proc. Natl Acad. Sci. USA* **72**, 2964.

Trimble, R. B. and Maley, F. (1977). *Biochem. Biophys. Res. Commun.* **78**, 935.

Tsuruo, T., Lida, H., Tsukayoshi, S. and Sakurai, Y. (1981). *Cancer Res.* **41**, 1967.

Tsuruo, T., Lida, H., Tsukayoshi, S. and Sakurai, Y. (1982). *Cancer Res.* **42**, 4730.

Tulsiani, D.R.P. and Carubelli, R. (1970). *J. Biol. Chem.* **245**, 1821.

Tuszynski, G.P., Baker, S.P., Fuhrer, J.P., Buck, C.A. and Warren, L. (1978). *J. Biol. Chem.* **253**, 6092.

Ueda, K., Cardarelli, C., Gottesmann, M.M. and Pastan, I. (1987). *Proc. Natl. Acad. Sci. USA* **84**, 3004.

Usuki, S., Hoops, P. and Sweeley, C.C. (1988) *J. Biol. Chem.* **263**, 10595.

van Beek, W.P., Smets, L.A. and Emmelot, P. (1975). *Nature* **253**, 457.

van der Bliek, A.M. and Borst, P. (1989). *Adv. Cancer Res.* **52**, 165.

Vimr, E.R., McCoy, R.D., Vollger, H.F., Wilkinson, N.C. and Troy, F.A. (1984). *Proc. Natl. Acad. Sci. USA* **81**, 1971.

Warner, T.G., Louie, A., Potier, M. and Ribiero, A. (1991). *Carbohydrate Res.* **215**, 315.

Warren, L. (1959). *J. Biol. Chem.* **234**, 1971.

Warren, L. (1960). *Biochim. Biophys. Acta* **44**, 347.

Warren, L. (1963). *Comp. Biochem. Physiol.* **10**, 153.

Warren, L. (1964). *Biochim. Biophys. Acta* **83**, 129.

Warren, L. (1989). *J. Biol. Chem.* **264**, 8835.

Warren, L. (1990). *Exp. Cell. Res.* **190**, 133

Warren, L. and Blacklow, R.S. (1962). *J. Biol. Chem.* **237**, 3527.

Warren, L. and Felsenfeld, H. (1962). *J. Biol. Chem.* **237**, 1421.

Warren, L. and Spearing, C.W. (1960). *Biochem. Biophys. Res. Commun.* **3**, 489.

Warren, L., Critchley, D. and MacPherson, I. (1972a). *Nature* (Lond.) **235**, 275.

Warren, L., Fuhrer, J.P. and Buck, C.A. (1972b). *Proc. Natl. Acad. Sci. USA* **69**, 1838.

Warren, L., Buck, C.A. and Tuszynski, G.P. (1978). *Biochim. Biophys. Acta* **516**, 97.

Warren, L., Blithe, D.L. and Cossu, G. (1982). *J. Cell Physiol.* **113**, 17.

Warren, L., Baker, S.R., Blithe, D.L. and Buck, C.A. (1983). In *Biomembranes*, Vol. 11, A. Nowotny, p. 53. New York: Plenum Press.

Warren, L., Jardillier, J.-C. and Ordentlich, P. (1991). *Cancer Res.* **51**, 1996.

Warren, L., Jardillier, J.-C., Malarska, A. and Akeli, M.-G. (1992). *Cancer Res.* **52**, 3241.

Wilson, G.M., Serrano, A.E., Wasley, A., Bogenschutz, M.P., Shankar, A.H. and Wirth, D.F. (1989). *Science* **244**, 1184.

Yamakawa, T. and Suzuki, S. (1952). *J. Biochem.* (Tokyo) **39**, 175.

Yamamoto, K., Tsuji, T., Tarutani, O. and Osawa, T. (1984). *Eur. J. Biochem.* **143**, 133.

Yamashita, K., Ohkura, T., Tachibana, Y., Takasaki, S. and Kobata, A. (1984). *J. Biol. Chem.* **259**, 10834.

Yamashita, K., Tachibana, Y., Ohkura, T. and Kobata, A. (1985). *J. Biol. Chem.* **260**, 3963.

Youakim, A., Romero, P.A., Yee, K., Carlsson, S.R., Fukuda, M. and Herscovics, A. (1989). *Cancer Res.* **49**, 6889.

Zamora, J.M., Pearce, H.L. and Beck, W.T. (1988). *Molec. Pharmacol.* **33**, 454.

Zapata, G., Vann, W.F., Aaronson, W., Lewis, M.S. and Moos, M. (1989). *J. Biol. Chem.* **264**, 14769.

Zhang, J.-T. and Ling, V. (1991). *J. Biol. Chem.* **266**, 18224.

Zhu, B.C.R. and Laine, R.A. (1985). *J. Biol. Chem.* **260**, 4041.

Index

Printed in the United States
By Bookmasters